Rock Fracture Mechanics and Fracture Criteria

Yu Zhao · Kun Zheng · Chaolin Wang

Rock Fracture Mechanics and Fracture Criteria

 Springer

Yu Zhao
College of Civil Engineering
Guizhou University
Guiyang, China

Kun Zheng
College of Civil Engineering
Guizhou University
Guiyang, China

Chaolin Wang
College of Civil Engineering
Guizhou University
Guiyang, China

ISBN 978-981-97-5821-0   ISBN 978-981-97-5822-7 (eBook)
https://doi.org/10.1007/978-981-97-5822-7

This work was supported by Guizhou University.

This Springer imprint is published by the registered company Springer Nature Singapore Pte Ltd.
The registered company address is: 152 Beach Road, #21-01/04 Gateway East, Singapore 189721, Singapore

If disposing of this product, please recycle the paper.

Preface

For actual rock masses with massive natural cracks, there are three basic fracture modes, including mode I (i.e., tension), mode II (i.e., shear), and mode III (i.e., tearing). For rock engineering projects (e.g., the cutting and fragmenting of rocks, tunnel excavation, shale gas and coalbed methane extraction, and an enhanced geothermal system [EGS] in hot dry rock [HDR]), exploring the fracture properties of rocks is helpful in achieving the efficient exploitation and improved safety. The fracture toughness of rocks is a crucial mechanical indicator, which weighs the stress and displacement fields near the crack front. When the stress intensity factor (SIF) reaches its critical level (i.e., the fracture toughness), the cracks become unstable and propagate rapidly. The fracture criterion can provide theoretical interpretations for the initiation, extension, and nucleation of cracks and is regarded as an important research issue in the discipline of fracture mechanics. To understand comprehensively the fracture characteristics of rocks, this book designs a new fracture test method and proposes some failure criteria.

This book first describes the significance of rock fracture mechanics and fracture criteria. Chapter 2 introduces the detailed derivations of established fracture criteria, which consider the singular and non-singular terms of stress fields near the crack tip. Chapter 3 reviews some popular fracture test methods, including mixed-mode I/II, true mode II, and mixed-mode I/III. Chapter 4 adopts the semi-circular bending (SCB) method to investigate the mixed-mode I/II fracture behaviors. Chapter 5 employs four different mode II specimens to explore the true mode II fracture properties, including the short core in compression (SCC), Z-shaped centrally cracked direct shear (ZCCDS), shear-box (SB), and punch-through shear (PTS). Chapter 6 uses the edge notched disk bending (ENDB) and double-edge notched disk compression (DENDC) specimens to study and compare the mode III loading and mode III fracturing problems.

This book is intended as a reference book for graduate students, engineers, and scientists in the fields of rock engineering, solid mechanics, and material science. The contents presented in this book are based on the research results of the three authors, which can provide an important reference for material design and stability analysis. During the writing of this book, we made abundant references to key publications in related fields and tried to tell the readers in the latest manner. Due to the limitation of our knowledge, there are some inevitable mistakes and defects in the book. Your suggestions would be deeply appreciated.

Guiyang, China   Yu Zhao
April 2024   Kun Zheng
                                                                                Chaolin Wang

Acknowledgements

This book is supported by the Research Fund for Talents of Guizhou University (Grant No. 201901), and Specialized Research Funds of Guizhou University (Grant No. 201903). We are grateful to the College of Civil Engineering, Guizhou University, which provides us an excellent work and research environment.

Our final thanks go to the publisher Springer Nature Singapore Pte Ltd. We acknowledge the excellent support of *Wayne Hu* (Publishing Editor) while working on the book manuscript.

Contents

Chapter 1
Introduction

For rock engineering projects, the cutting and fragmenting of rocks has attracted much attention. Exploring the fracture characteristics of rocks is helpful in achieving efficient and sustainable excavation for mining and tunneling engineering. The fracture toughness of rocks is a crucial mechanical indicator that weighs the initiation and extension of cracks during the breaking process. The microwave-assisted rock-fragmentation approach, as a clean, efficient, and convenient method of rock fragmentation, was adopted by Yang et al. [1] to investigate combined-mode I/II fracture properties of basalt under different irradiation times ranging from 0 to 300 s. An EGS (enhanced geothermal system) can effectively develop geothermal resources in HDR (hot dry rock). The HDR reservoir with massive fracture networks is expected to increase the contact regions between the injected water and rock masses and improve the geothermal output. Therefore, it is essential to understand the fracture mechanical properties of HDR reservoirs. Since granite is regarded as a typical rock type in geothermal energy extraction, Feng et al. [2] investigated the thermal effect on combined-mode I/II fracture behaviors of granite under different temperatures ranging from 20 °C to 600 °C. Shale rocks are characterized by extremely low porosity and permeability, and the abundant shale gas can be successfully extracted by the fracking (or hydraulic fracturing) technique. The fracking technique enhances the productivity and recovery of shale gas by creating high-conductivity fractures, requiring a thorough comprehension of fracture network formation. When the SIF (stress intensity factor) attains its critical state, the cracks become unstable and propagate rapidly, and the critical SIF can significantly affect the geometry of hydraulic fractures. As a central fracturing index, shale fracture toughness weighs the stress and displacement fields near the crack front, and controls the formation and distribution of hydraulic fracturing networks. Currently, the investigators have emphasized on the evaluation of pure mode I fracture resistance for shale via the ENRBB (edge-notched rectangular beam bending) [3–5], ENSCB (edge-notched semicircular bending) [6–8], CCDC (centrally cracked disk compression) [9], and CCNDC (centrally chevron-notched disk compression) [10] testing methods, while the true mode-II fracture resistance of bedded shale has been effectively estimated via three shear approaches [11–13]. In fact, naturally cracked reservoirs generally undergo the complicated combined-mode I/II loading, and the resultant fracturing problems were analyzed by Suo et al. [14] and Wang et al. [15] using the ENSCB and HCCD (hollow center cracked disk) testing methods.

For actual rock masses with randomly internal cracks, there are three essential fracture types, namely pure mode I (tension), pure mode II (shear), and pure mode III (tearing) which generate the opening, planar sliding, and nonplanar sliding deformations of cracks respectively. In the fields of solid fracture mechanics and hydraulic fracturing, fracture toughness is referred to as an important parameter for evaluating the resistance of engineering materials to cracking growth [16–20]. To determine the magnitude of fracture toughness, different test specimens and loading configurations are devised and customized depending on engineering requirements [21–31]. Considering the economy and convenience of fracture testing, the SCB (semi-circular bending) specimen was designed by Chong and Kuruppu [32] as originators who concluded that this specimen can be readily manufactured from rock masses and is loaded by a common and straightforward three-point bending fixture. Under different SCB specimen sizes, the variations of mode-I fracture resistance could be successfully predicted by a stress-based fracture model [33]. Compared to another SCB specimen with a chevron-shaped notch, the SCB specimen with a straight notch is an expedient measure for the assessment of mode-I fracture resistance [34]. Awaj and Sato [35] and Atkinson et al. [36] were among the first investigators who adopted the CCBD (centrally cracked Brazilian disk) specimen to determine the dimensionless geometry factors via numerical and theoretical approaches. Using the popular CCBD specimen, the mode-I fracturing behaviors under different engineering factors (e.g., drying and watering cycle [37–39], microwave treatment [40], loading rate [41], and bedding direction [42]) were reported in succession. For the SECRBB (single-edge cracked round bar bending) specimen with a span to diameter ratio 3.3, a wide range of expressions for nondimensional mode-I SIF (stress intensity factor) were acquired from experimental and analytical results [43, 44]. Due to the complexity of conventional computation methods [45], the comprehensive and practical formulas for the CNBB (chevron-notched beam bending) specimen were proposed to facilitate the determination of mode-I fracture resistance [46]. To reduce test costs and material consumption, Aliha et al. [47] invented the SENSBB (single-edge notched short beam bending) specimen instead of the conventional SENBB (single-edge notched beam bending) specimen [48]. Moreover, the gamut of mode I/II fracture resistance can be estimated by the improved SENSBB specimen [47]. In addition, the ECTB, ENDB, ENDC, and HCCD specimens [49–52] were developed to provide the fracture parameters available for specific engineering applications. Particularly, the above-mentioned test specimens can produce the complete range of mode mixities from pure mode I (tension) to pure mode II (shear) or III (tearing) depending on the geometry and loading conditions of test specimens [53–56].

The fracture criterion can provide theoretical interpretations for the initiation, extension, and nucleation of cracks and is regarded as an important research issue in the discipline of fracture mechanics. To theoretically forecast the onset of fractrue, numerous fracture criteria were proposed and developed from the perspectives of stress, strain, and energy. In 1921, the first fracture law for cracking development was built by Griffth [56] as an originator who stated that the fracture is imminent when the accumulative energy in solids reached the energy consumed to form new fracturing surfaces. The K-criterion is created by Irwin [57] in 1957 referring to the concept of SIF (stress intensity factor). It assumed that the initial fracturing started to happen when the K_I (mode-I SIF) attained the K_{Ic} (mode-I fracture toughness) of materials. Noticing that the aforementioned two criteria are merely defined under the pure mode-I fracturing [58]. In 1963, the 2D-MTS (maximum tangential stress) criterion was established by Erdogan and Sih [59] as the pioneers who considered the singular terms and neglected the nonsingular terms in Willimas' stress solution [60]. Subsequently, Liu et al. [61] extended the 2D-MTS criterion to the establishment of the 3D-MTS criterion in 2004. According to the elastic theory of Hooke, the 2D-MTSN (maximum tangential strain) criterion of the plane stress state was presented by Chang [62] in 1981. Then the 3D-MTSN criterion was developed by Bidadi et al. [63] in 2022 to make it suitable for 3D fracturing problems. According to the definitions of total strain energy density and tangential strain energy density, Sih [64] and Koo and Choy [65] respectively found the 2D-MSEDF (minimum strain energy density factor) and 2D-MTSEDF (maximum tangential strain energy density factor) criteria. Later, Sih [66] and Ayatollahi et al. [67] respectively developed the 3D-MSEDF and 3D-MTSEDF criteria to make it applicable to 3D fracturing problems. Based on the definition of SIF and the hypothesis of continuum mechanics, 2D-MERR (maximum energy release rate) was presented by Nuismer [68] in 1975. According to the correlation between ERR and SIF, the 3D-MERR criterion was derived by Zhao [7] in 1987 to make it appropriate for the 3D fracturing analyses. Since the rigorous mathematical form of ERR in the arbitrary branching crack direction is almost impossible to be determined, the 3D-MPERR (maximum potential energy release rate) criterion was suggested by Chang et al. [70] in 2006 to approximatively interpret the 3D fracturing problems. Supposing that the plastic boundary around a cracking tip is limited and the total strain energy is responsible for the fracture initiation, the 3D-MSE (maximum strain energy) or 3D-MPZR (minimum plastic zone radius) criterion was presented by Zhao [71] in 1987. The stress-based criteria includ MTS (maximum tangential stress), MPS (maximum principal stress) [72], MES (maximum effective stress) [73], and MMPS (maximum mean principal stress) [74]. The strain-based criteria include MTSN (maximum tangential strain), MPSN (maximum principal strain) [75], and MPSN-ECD (maximum principal strain considering an effective critical radius or distance) [76]. While the energy-based criteria include MERR (maximum energy release rate), ASED (average strain energy density) [77], MTSEDF (maximum tangential strain energy density factor), MSEDF (minimum strain energy density factor). As these conventional fracture criteria are inadequate in predicting the envelope of combined-mode I/II fracture resistance, they have been modified and extended by incorporating the non-singular T-stress term. The modified fracture models can provide acceptable and satisfactory estimations for the experimental data, this indicates that the constant T-stress plays an important role in enhancing the accuracy of predicting brittle fracture [79–88].

The selection of appropriate test specimens and fracture criteria is of significance to investigate the fracture properties of distinct engineering materials, and they should be modified and optimized according to specific engineering and experiment requirements. On the other hand, the fracture mechanisms are expected to obtain for the thorough comprehension of material properties via high-performance monitoring techniques [89–100], for instance, AE (acoustic emission), CT (X-ray computed tomography), and NMR (nuclear magnetic resonance). Particularly, the AE technique has been widely applied by scientists and engineers in experiment study, engineering health assessment, and rockburst forewarning because it possesses a series of advantages: real-time, efficient, non-destructive, and economical [101–106]. Generally, the emphasis of AE technique applications is to utilize the AE signal-based analysis approaches for revealing the mechanical mechanisms of materials and structures. There are three analysis approaches: AE signal positioning, AE parameter analysis, and AE spectrum analysis. Note that the precision of AE signal positioning depends heavily on the AE sensor alignment and wave velocity adjustment. Further, the engineering applications of AE signal positioning could be limited. Due to the multiscale and multidimensional nature of microcrack growth, classical AE parameter analysis cannot fully capture its complexity. Hence, the conventional RA-AF analysis of AE signals is compared and discussed via improved methods: Bayesian moment tensor inversion algorithm [105], numerical simulation [106], Gaussian mixture model [107], tension and shear tests [108], AF-IF analysis [109], and RA-AF-based Kernel density median standard [110]. With the development of high-tech computing processors, the AE spectrum analysis can promote a novel potential in mechanical mechanism discernment [111]. This is because cracking growth incidents have distinct AE spectrum signatures depending on failure modes (e.g., shearing and tension) [112].

References

1. Yang Z, Yin T, Wu Y, Zhuang D, Yin J, Ma J (2023) Mixed-mode I/II fracture properties and failure characteristics of microwave-irradiated basalt: an experimental study. Fatigue Fract Eng Mater Struct
2. Feng G, Zhu C, Wang X, Tang S (2023) Thermal effects on prediction accuracy of dense granite mechanical behaviors using modified maximum tangential stress criterion. J. Rock Mech. Geotech. Eng. 15(7):1734–1748
3. Dou FK, Wang JG, Zhang XX, Wang HM (2019) Effect of joint parameters on fracturing behavior of shale in notched three-point-bending test based on discrete element model. Eng Fract Mech 205:40–56
4. Liu J, Yao K, Xue Y, Zhang XX, Chong ZH, Liang X (2019) Study on fracture behavior of bedded shale in three-point-bending test based on hybrid phase-field modelling. Theor Appl Fract Mec 104:102382
5. Li CB, Yang DC, Xie HP, Ren L, Wang J (2021) Research on the anisotropic fracture behavior and the corresponding fracture surface roughness of shale. Eng Fract Mech 255:107963
6. Li YJ, Wang ST, Zheng LG, Zhao SK, Zuo JP (2021) Evaluation of the fracture mechanisms and criteria of bedding shale based on three-point bending experiment. Eng Fract Mech 255:107913

7. Shi XS, Zhao YX, Gong S, Wang W, Yao W (2022) Co-effects of bedding planes and loading condition on mode-I fracture toughness of anisotropic rocks. Theor Appl Fract Mec 117:103158

8. Lei B, Li HT, Zuo JP, Liu HY, Yu ML, Wu GS (2021) Meso-fracture mechanism of Longmaxi shale with different crack-depth ratios: Experimental and numerical investigations. Eng Fract Mech 257:108025

9. Xie Q, Liu XL, Li SX, Du K, Gong FQ, Li XB (2022) Prediction of mode I fracture toughness of shale specimens by different fracture theories considering size effect. Rock Mech Rock Eng 55:7289–7306

10. Shi XS, Zhao YX, Danesh NN, Zhang X, Tang TW (2022) Role of bedding plane in the relationship between mode-I fracture toughness and tensile strength of shale, B. Eng Geol Environ 81:81

11. Fan ZD, Xie HP, Ren L, Zhang R, He R, Li CB, Zhang ZT, Wang J, Xie J (2022) Anisotropy in shear-sliding fracture behavior of layered shale under different normal stress conditions. J Cent South Univ 29(11):3678–3694

12. Fan ZD, Xie HP, Zhang R, Lu HJ, Zhou Q, Nie XF, Luo Y, Ren L (2022) Characterization of anisotropic mode II fracture behaviors of a typical layered rock combining AE and DIC techniques. Eng Fract Mech 271:108599

13. Sun DL, Rao QH, Wang SY, Shen QQ, Yi W (2021) Shear fracture (Mode II) toughness measurement of anisotropic rock. Theor Appl Fract Mec 115:103043

14. Suo Y, Su XH, Ye QY, Chen ZX, Feng FP, Wang XY, Xie K (2022) The investigation of impact of temperature on mixed-mode fracture toughness of shale by semi-circular bend test. J Petrol Sci Eng 217:110905

15. Wang H, Li Y, Cao SG, Fantuzzi N, Pan RK, Tian MY, Liu YB, Yang HY (2020) Fracture toughness analysis of HCCD specimens of Longmaxi shale subjected to mixed mode I-II loading. Eng Fract Mec 239:107299

16. Wang CL, Zhao Y, Ning L, Bi J (2022) Permeability evolution of coal subjected to triaxial compression based on in-situ nuclear magnetic resonance. Int J Rock Mech Min Sci 159:105213

17. Zhao Y, Wang CL, Ning L, Zhao HF, Bi J (2022) Pore and fracture development in coal under stress conditions based on nuclear magnetic resonance and fractal theory. Fuel 309:122112

18. Guo TF, Liu KW, Li X, Wu Y, Yang JC (2023) Effects of thermal treatment on the fracture behavior of rock-concrete bi-material specimens containing an interface crack. Theor Appl Fract Mech 127:104071

19. Zhang YF, Long AF, Zhao Y, Zang A, Wang CL (2023) Mutual impact of true triaxial stress, borehole orientation and bedding inclination on laboratory hydraulic fracturing of Lushan shale, J Rock Mech Geotech

20. Wei MD, Dai F, Liu Y, Jiang RC (2023) A fracture model for assessing tensile mode crack growth resistance of rocks. J Rock Mech Geotech Eng 15:395–411

21. Moghaddam MR, Ayatollahi MR, Berto F (2017) Mixed mode fracture analysis using generalized averaged strain energy density criterion for linear elastic materials. Int J Solids Struct 120:137–145

22. Rashidi Moghaddam M, Ayatollahi MR, Berto F (2018) Rock fracture toughness under mode II loading: a theoretical model based on local strain energy density, Rock Mech Rock Eng 51:243–253

23. Ayatollahi M, Saboori B (2015) Maximum tangential strain energy density criterion for general mixed mode I/II/III brittle fracture. Int J Damage Mech 24:263–278

24. Ayatollahi MR, Saboori B (2015) T-stress effects in mixed mode I/II/III brittle fracture. Eng Fract Mech 144:32–45

25. Karimi HR, Aliha MRM, Ebneabbasi P, Salehi SM, Khedri E, Haghighatpour PJ (2023) Mode I and mode II fracture toughness and fracture energy of cement concrete containing different percentages of coarse and fine recycled tire rubber granules. Theor Appl Fract Mech 123:103722

26. Khansari NM, Aliha MRM (2023) Mixed-modes (I/III) fracture of aluminum foam based on micromechanics of damage. Int J Damage Mech 32(4):519–548

27. Omidvar N, Aliha MRM, Khoramishad H (2023) Hygrothermal degradation of MWCNT/epoxy brittle materials under I/II combined mode loading conditions: an experimental, micro structural and theoretical study. Theor Appl Fract Mech 125:103896

28. Mehraban MR, Bahrami B, Ayatollahi MR, Nejati M (2023) A non-local XFEM-based methodology for modeling mixed-mode fracturing of anisotropic rocks. Rock Mech Rock Eng 56:895–909

29. Bahrami B, Talebi H, Ayatollahi MR, Khosravani MR (2023) Artificial neural network in prediction of mixed-mode I/II fracture load. Int J Mech Sci 248:108214

30. Sapora A, Ferrian F, Cornetti P, Talebi H, Ayatollahi MR (2023) Ligament size effect in largely cracked tensile structures. Theor Appl Fract Mech 125:103871

31. Bahrami B, Ayatollahi MR, Mehraban MR, Nejati M, Berto F (2022) On the effects of higher order stress terms in pure mode III loading of bi-material notches. Fatigue Fract Eng Mater Struct 45:3333–3346

32. Chong KP, Kuruppu MD (1984) New specimen for fracture toughness determination for rock and other materials. Int J Fracture 26:59–62

33. Ayatollahi MR, Akbardoost J (2013) Size effects in mode II brittle fracture of rocks. Eng Fract Mech 112–113:165–180

34. Obara YZ, Nakamura K, Yoshioka S, Sainoki A, Kasai A (2019) Crack front geometry and stress intensity factor of semi-circular bend specimens with straight through and chevron notches. Rock Mech Rock Eng

35. Awaji H, Sato S (1978) Combined mode fracture toughness measurement by the disc test. J Eng Mater Technol 100:175–182

36. Atkinson C, Smelser RE, Sanchez J (1982) Combined mode fracture via the cracked Brazilian disk test. Int J Fract 18:279–291

37. Hua W, Dong SM, Li YF, Xu JG, Wang QY (2015) The influence of cyclic wetting and drying on the fracture toughness of sandstone. Int J Rock Mech Min 78:331–335

38. Hua W, Dong SM, Li YF, Wang QY (2016) Effect of cyclic wetting and drying on the pure mode II fracture toughness of sandstone. Eng Fract Mech 153:143–150

39. Hua W, Dong SM, Peng F, Li KY, Wang QY (2017) Experimental investigation on the effect of wetting-drying cycles on mixed mode fracture toughness of sandstone. Int J Rock Mech Min 93:242–249

40. Ge ZL, Sun Q, Xue L, Yang T (2021) The influence of microwave treatment on the mode I fracture toughness of granite. Eng Fract Mech 249:107768

41. Liu J, Qiao L, Li Y, Li QW, Fan DJ (2022) Experimental study on the quasi-static loading rate dependency of mixed-mode I/II fractures for marble rocks. Theor Appl Fract Mech 121:103431

42. Zheng K, Wang CL, Zhao Y, Bi J (2023) Theoretical and experimental researches on fracture toughness for bedded shale using the centrally cracked Brazilian disk method with acoustic emission monitoring. Theor Appl Fract Mech 124:103784

43. Bush AJ (1976) Experimentally determined stress-intensity factors for single-edge-crack round bars loaded in bending. Exp mech 16:249–257

44. Underwood JH, Woodward RL (1989) Wide range stress-intensity-factor expression for an edge-cracked round bar bend specimen. Exp Mech 29:166–168

45. Wu SX (1984) Fracture toughness determination of bearing steel using chevron-notch three point bend specimen. Eng Fract Mech 19:221–232

46. Deng X, Bitler J, Chawla KK, Patterson BR (2010) Toughness measurement of cemented carbides with chevron-notched three-point bend test. Adv Eng Mater 12

47. Aliha MRM, Samareh-Mousavi SS, Mirsayar MM (2021) Loading rate effect on mixed mode I/II brittle fracture behavior of PMMA using inclined cracked SBB specimen. Int J Solids Struct 232:111177

48. Zhao Y, Zheng K, Wang CL, Bi J, Zhang H (2022) Investigation on model-I fracture toughness of sandstone with the structure of typical bedding inclination angles subjected to three-point bending. Theor Appl Fract Mech 119:103327

49. Aliha MRM, Hosseinpour GhR, Ayatollahi MR (2013) Application of cracked triangular specimen subjected to three-point bending for investigating fracture behavior of rock materials. Rock Mech Rock Eng 46:1023–1034

50. Pietras D, Aliha MRM, Kucheki HG, Sadowski T (2023) Tensile and tear-type fracture toughness of gypsum material: direct and indirect testing methods. J Rock Mech Geotech Eng 15:1777–1796

51. Bahmani A, Farahmand F, Janbaz MR, Darbandi AH, Ghesmati-Kucheki H, Aliha MRM (2021) On the comparison of two mixed-mode I + III fracture test specimens. Eng Fract Mech 241:107434

52. Karimi HR, Bidadi J, Aliha MRM, Mousavi A (2023) Mohammadi MH, Haghighatpour PJ (2023) An experimental study and theoretical evaluation on the effect of specimen geometry and loading configuration on recorded fracture toughness of brittle construction materials. J Build Eng 75:106759

53. Jalayer R, Saboori B, Ayatollahi MR (2023) A novel test specimen for mixed mode I/II/III fracture study in brittle materials. Fatigue Fract Eng Mater Struct 1–13

54. Zhao YX, Sun Z, Gao YR, Wang XL, Song HH (2022) Influence of bedding planes on fracture characteristics of coal under mode II loading. Theor Appl Fract Mech 117:103131

55. Wang W, Teng T (2022) Experimental study on anisotropic fracture characteristics of coal using notched semi-circular bend specimen. Theor Appl Fract Mec 122:103559

56. Griffith AA (1921) The phenomena of rupture and flow in solids. Philos Trans R Soc Lond 221:163–198

57. Irwin GR (1957) Analysis of stresses and strains near end of a crack traversing a plate. J Appl Mech 24:361–364

58. Hua W, Li JX, Zhu ZY, Li AQ, Huang JZ, Gan ZQ, Dong SM (2023) A review of mixed mode I-II fracture criteria and their applications in brittle or quasi-brittle fracture analysis. Theor Appl Fract Mec 124:103741

59. Erdogan F, Sih GC (1963) On the crack extension in plates under plane loading and transverse shear. J Basic Eng 85:519–525

60. Williams ML (1957) On the stress distribution at the base of a stationary crack. ASME J Appl Mech 24:109–114

61. Liu S, Chao YJ, Zhu X (2004) Tensile-shear transition in mixed mode I/III fracture. Int J Solids Struct 41(2004):6147–6172

62. Chang KJ (1981) On the maximum strain criterion—a new approach to the angled crack problem. Eng Fract Mech 14:107–124

63. Bidadi J, Aliha MRM, Akbardoost J (2022) Development of maximum tangential strain (MTSN) criterion for prediction of mixed-mode I/III brittle fracture. Int J Solids Struct 256:111979

64. Sih GC (1974) Strain-energy-density factor applied to mixed mode crack problems. Int J Fract 10:305–321

65. Koo JM, Choy YS (1991) A new mixed mode fracture criterion: maximum tangential strain energy density criterion. Eng Fract Mech 39:443–449

66. Sih G (1991) A three-dimensional strain energy density factor theory of crack propagation. Springer, Mechanics of Fracture Initiation and Propagation, pp 23–56

67. Ayatollahi M, Saboori B (2015) Maximum tangential strain energy density criterion for general mixed mode I/II/III brittle fracture. Int J Damage Mech 24:263–278

68. Nuismer R (1975) An energy release rate criterion for mixed mode fracture. Int J Fract 11:245–250

69. Zhao Y (1987) Griffith's criterion for mixed mode crack propagation. Eng Fract Mech 26:683–689

70. Chang J, Xu JQ, Mutoh Y (2006) A general mixed-mode brittle fracture criterion for cracked materials. Eng Fract Mech 73:1249–1263

71. Zhao Y (1987) A strain energy criterion for mixed mode crack propagation. Eng Fract Mech 26:533–539

72. Schöllmann M, Richard HA, Kullmer G, Fulland M (2002) A new criterion for the prediction of crack development in multiaxially loaded structures. Int J Fract 117:129–141

73. Sajjadi SH, Ostad Ahmad GHorabi MJ, Salimi-Majd D (2015) A novel mixed-mode brittle fracture criterion for crack growth path prediction under static and fatigue loading. Fatigue Fract Eng Mater Struct 38:1372–1382

74. Wang J, Ren L, Xie LZ, Xie HP, Ai T (2016) Maximum mean principal stress criterion for three-dimensional brittle fracture. Int J Solids Struct 102–103:142–154

75. Fischer KF, Göldner H (1981) On the formulation of a principal strain criterion in crack fracture mechanics. Int J Fract 17:R3–R6

76. Mirsayar MM (2021) On the effective critical distances in three-dimensional brittle fracture via a strain-based framework. Eng Fract Mech 248:107740

77. Moghaddam MR, Ayatollahi MR, Berto F (2017) Mixed mode fracture analysis using generalized averaged strain energy density criterion for linear elastic materials. Int J Solids truct. 120:137–145

78. Smith DJ, Ayatollahi MR, Pavier MJ (2001) The role of T-stress in brittle fracture for linear elastic materials under mixed-mode loading. Fatigue Fact Eng Mater Struct 24:137–150

79. Hou C, Jin XC, Fan XL, Xu R, Wang ZY (2019) A generalized maximum energy release rate criterion for mixed mode fracture analysis of brittle and quasi-brittle materials. Theor Appl Fract Mech 100:78–85

80. Shen Z, Yu HY, Guo LC, Hao LL, Zhu S, Huang K (2023) A modified 3D G-criterion for the prediction of crack propagation under mixed mode I-III loadings. Eng Fract Mech 281:109082

81. Hua W, Li JX, Zhu ZY, Li AQ, Huang JZ, Gan ZQ, Dong SM (2023) A review of mixed mode I-II fracture criteria and their applications in brittle or quasi-brittle fracture analysis. Theor Appl Fract Mech 124:103741

82. Hua W, Li JX, Zhu ZY, Li AQ, Huang JZ, Dong SM (2023) Experimental study on mode I and mode II fracture properties of heated sandstone after two different cooling treatments. Geomech Energy Environ 100448

83. Yang Z, Yin TB, Wu Y, Zhuang DD, Yin JW, Ma JX (2023) Mixed-mode I/II fracture properties and failure characteristics of microwave-irradiated basalt: an experimental study. Fatigue Fract Eng Mater Struct 46:814–834

84. Aliha MRM, Ayatollahi MR, Akbardoost J (2012) Typical upper bound–lower bound mixed mode fracture resistance envelopes for rock material. Rock Mech Rock Eng 45:65–74

85. Aliha MRM, Mahdavi E, Ayatollahi MR (2016) The influence of specimen type on tensile fracture toughness of rock materials. Pure Appl Geophys

86. Aliha MRM, Mahdavi E, Ayatollahi MR (2018) Statistical analysis of rock fracture toughness data obtained from different chevron notched and straight cracked mode I specimens. Rock Mech Rock Eng

87. Yu J, Zhu YL, Yao W (2021) Stress relaxation behaviour of marble under cyclic weak disturbance and confining pressures. Measurement 182

88. Yu J, Yao W, Duan K, Liu XY, Zhu YL (2020) Experimental study and discrete element method modeling of compression and permeability behaviors of weakly anisotropic sandstones. Int J Rock Mech Min 134:104437

89. Bi J, Ning L, Zhao Y, Wu ZJ, Wang CL (2023) Analysis of the microscopic evolution of rock damage based on real-time nuclear magnetic resonance. Rock Mech Rock Eng 56:3399–3411

90. Bi J, Tang JC, Wang CL, Quan DG, Teng MY (2022) Crack coalescence behavior of rock-like specimens containing two circular embedded flaws. Lithosphere (Special 11) 9498148

91. Bi J, Liu PF, Gan F (2020) Effects of the cooling treatment on the dynamic behavior of ordinary concrete exposed to high temperatures. Constr Build Mater 248:118688

92. Wang Y, Yi XF, Han JQ, Xia YJ (2022) Acoustic emission and computed tomography investigation on fatigue failure of fissure-contained hollow-cylinder granite: cavity diameter effect. Fatigue Fract Eng Mater Struct 45:2243–2260

93. Niu Y, Wang JG, Wang XK, Hu YJ, Zhang JZ, Zhang RR, Hu ZJ (2023) Numerical study on cracking behaviors and fracture failure mechanism of flawed rock materials under uniaxial compression. Fatigue Fract Eng Mater Struct 46(6):2096−2111

94. Niu Y, Liu PF, Zhang CP, Hu YJ, Wang JG (2023) Mechanical properties and dynamic multifractal characteristics of shale under anisotropic stress using AE technology. Geoenergy Sci Eng 226:211748

95. Niu Y, Wang G, Wang JG, Liu XQ, Zhang RR, Qiao JX, Zhang JZ (2023) Experimental study on thermal fatigue damage and failure mechanisms of basalt exposed to high-temperature Treatments. Fatigue Fract Eng Mater Struct 1–20

96. Zhou XP, Peng SL, Zhang JZ, Zhou JN, Berto F (2022) Experimental study on cracking behaviors of coarse and fine sandstone containing two flaws under biaxial compression. Fatigue Fract Eng Mater Struct 45:2595–2612

97. Zhou XP, Huo ZL, Berto F (2022) Stick–slip shear failure along bimaterial interfaces: an experimental study on granite and basalt. Fatigue Fract Eng Mater Struct 45:2023–2046

98. Wu K, Meng QS, Wang C, Qin QL, Dong ZW (2023) Investigation of damage characteristics of coral reef limestone under uniaxial compression based on pore structure. Eng Geol 313:106976

99. Teng MY, Bi J, Zhao Y, Wang CL (2023) Experimental study on shear failure modes and acoustic emission characteristics of rock-like materials containing embedded 3D flaw. Theor Appl Fract Mec 124:103750

100. Zhao Y, Lian SL, Bi J, Wang CL, Zheng K (2022) Study on freezing-thawing damage mechanism and evolution model of concrete. Theor Appl Fract Mec 121:103439

101. Wang CL, Zhao Y, He C, Bi J (2022) Study on the tensile-shear mechanical behavior of sandstone using a simple auxiliary apparatus. Theor Appl Fract Mec 122:103608

102. Zhang KP, Wang CL, Zhao Y, Bi J (2023) Experimental study on cracking behavior of concrete containing hole defects. J Build Eng 65:105806

103. Chen Y, Xu J, Peng SJ, Jiao F, Chen CC, Xiao ZY (2021) Experimental study on the acoustic emission and fracture propagation characteristics of sandstone with variable angle joints. Eng Geol 292:106247

104. Li SJ, Yang DX, Huang Z, Gu QX, Zhao K (2022) Acoustic emission characteristics and failure mode analysis of rock failure under complex stress state. Theor Appl Fract Mec 122:103666

105. Dong LJ, Zhang YH, Bi SJ, Ma J, Yan YH, Cao H (2023) Uncertainty investigation for the classification of rock micro-fracture types using acoustic emission parameters Int. J Rock Mech Min Sci 162:105292

106. Zhao K, Yang DX, Zeng P, Huang Z, Wu WK, Li B, Teng TY (2021) Effect of water content on the failure pattern and acoustic emission characteristics of red sandstone. Int J Rock Mech Min Sci 142:104709

107. Ju SY, Li DS, Jia JQ (2022) Machine-learning-based methods for crack classification using acoustic emission technique. Mech Syst Signal Pr 178:109253

108. Du K, Li XF, Tao M, Wang SF (2020) Experimental study on acoustic emission (AE) characteristics and crack classification during rock fracture in several basic lab tests. Int J Rock Mech Min Sci 133:104411

109. Rodríguez P, Celestino TB (2019) Application of acoustic emission monitoring and signal analysis to the qualitative and quantitative characterization of the fracturing process in rocks. Eng Fract Mech 210:54–69

110. Zhang JZ, Zhou XP (2023) Integrated acoustic-optic-mechanics (AOM) multi-physics field characterization methods for a crack: tension vs. shear. Eng Fract Mech 109339

111. Niu Y, Zhou XP, Berto F (2020) Temporal dominant frequency evolution characteristics during the fracture process of flawed red sandstone. Theor Appl Fract Mech 110:102838

112. Aggelis DG (2011) Classification of cracking mode in concrete by acoustic emission parameters. Mech Res Commun 38(3):153–157

Chapter 2
Review of Fracture Criteria

Under the combined-mode I/II loading, by taking into account the non-singular and singular terms of the solution established by Williams [1], the stress field at the adjacency of a cracking cusp in the polar coordinate system (see Fig. 2.1 [2]) can be denoted by the following formulas:

$$
\begin{cases}
\sigma_{rr} = \dfrac{1}{\sqrt{2\pi r}} \cos\dfrac{\theta}{2}\left[K_I(1+\sin^2\dfrac{\theta}{2}) + K_{II}(\dfrac{3}{2}\sin\theta - 2\tan\dfrac{\theta}{2})\right] + T\cos^2\theta + O(r^{1/2}) \\
\sigma_{\theta\theta} = \dfrac{1}{\sqrt{2\pi r}} \cos\dfrac{\theta}{2}(K_I\cos^2\dfrac{\theta}{2} - \dfrac{3}{2}K_{II}\sin\theta) + T\sin^2\theta + O(r^{1/2}) \\
\tau_{r\theta} = \dfrac{1}{2\sqrt{2\pi r}} \cos\dfrac{\theta}{2}[K_I\sin\theta + K_{II}(3\cos\theta - 1)] - T\sin\theta\cos\theta + O(r^{1/2})
\end{cases}
$$

$$(2.1)$$

where r and θ refer to the cracking tip coordinates in the polar system, σ_{rr}, $\sigma_{\theta\theta}$, and $\tau_{r\theta}$ signify the stress components, the first non-singular term T represents the constant T-stress parallel to the cracking orientation, $O(r^{1/2})$ means the higher order term and can be generally neglected near the cracking tip.

2.1 Maximum Tangential Stress (MTS) Criterion

According to the improved criterion proposed by Smith et al. [3], the fracture initiation angle θ_c corresponding to the critical tangential stress $\sigma_{\theta\theta c}$ should be satisfied as

$$
\frac{\partial \sigma_{\theta\theta}}{\partial \theta}\Big|_{\theta=\theta_c} = 0 \text{ and } \frac{\partial^2 \sigma_{\theta\theta}}{\partial^2 \theta}\Big|_{\theta=\theta_c} < 0 \tag{2.2}
$$

© The Author(s) 2024
Y. Zhao et al., *Rock Fracture Mechanics and Fracture Criteria*,
https://doi.org/10.1007/978-981-97-5822-7_2

Fig. 2.1 Stress field near the tip of a sharp crack in polar coordinate system

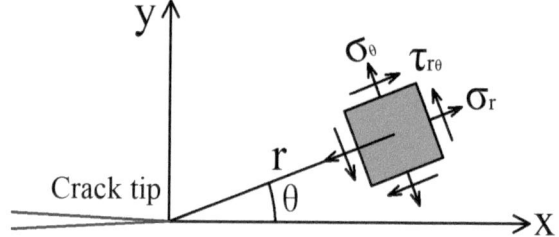

Based on the above-mentioned hypothesis, substituting Eq. (2.1) into Eq. (2.2) leads to

$$K_I \sin \theta_c + K_{II}(3 \cos \theta_c - 1) - \frac{16}{3} T \sqrt{2\pi r_c} \cos \theta_c \sin \frac{\theta_c}{2} = 0 \qquad (2.3)$$

where r_c is the critical damaged radius from the preexisting cracking tip. Obviously, the angle θ_c can be determined from Eq. (2.3).

In addition, the beginning of fracturing extension happens when the tangential stress $\sigma_{\theta\theta}$ along θ_c and at r_c overcomes its critical value $\sigma_{\theta\theta c}$, then Eq. (2.1) is modifed as

$$\sigma_{\theta\theta}(r_c, \theta_c) = \sigma_{\theta\theta c} = \frac{1}{\sqrt{2\pi r_c}} \cos \frac{\theta_c}{2} (K_I \cos^2 \frac{\theta_c}{2} - \frac{3}{2} K_{II} \sin \theta_c) + T \sin^2 \theta_c \quad (2.4)$$

Under the pure mode-I fracture conditions, there are $K_{II} = 0$, $K_I = K_{Ic}$, and $\theta_c = 0$, then Eq. (2.4) can be converted to

$$\sqrt{2\pi r_c} \sigma_{\theta\theta c} = K_{Ic} \qquad (2.5)$$

where K_{Ic} refers to the critical mode-I SIF and is regarded as the mode-I fracture toughness pertaining to the fracturing load P_{max}. Introducing Eq. (2.5) into Eq. (2.1) yields

$$K_{Ic} = \cos \frac{\theta_c}{2} (K_I \cos^2 \frac{\theta_c}{2} - \frac{3}{2} K_{II} \sin \theta_c) + T \sqrt{2\pi r_c} \sin^2 \theta_c \qquad (2.6)$$

Consequently, the envelopes for theoretical fracture resistance can be portrayed by the fracture resistance ratios K_I/K_{Ic} and K_{II}/K_{Ic} extracted from Eq. (2.6) in the normalized forms of F_I, F_{II}, and T^* as follows:

$$\begin{cases} \dfrac{K_I}{K_{Ic}} = \left[\cos \dfrac{\theta_c}{2} (\cos^2 \dfrac{\theta_c}{2} - \dfrac{3}{2} \dfrac{K_{11}}{K_I} \sin \theta_c) + \sqrt{2\pi r_c} \dfrac{T}{K_I} \sin^2 \theta_c \right]^{-1} \\[4mm] \dfrac{K_{II}}{K_{Ic}} = \left[\cos \dfrac{\theta_c}{2} (\dfrac{K_1}{K_{II}} \cos^2 \dfrac{\theta_c}{2} - \dfrac{3}{2} \sin \theta_c) + \sqrt{2\pi r_c} \dfrac{T}{K_{II}} \sin^2 \theta_c \right]^{-1} \end{cases} \qquad (2.7)$$

2.2 Maximum Tangential Strain (MTSN) Criterion

In compliance with the elastic theory of Hooke, the tangential strain $\varepsilon_{\theta\theta}$ is formulated for the special plane problems as follows [4, 5]:

$$\varepsilon_{\theta\theta} = \frac{1+\nu}{E}[\kappa\sigma_{\theta\theta} + (\kappa - 1)\sigma_{rr}] \tag{2.8}$$

where E and ν mean the Young's modulus and the Poisson's coefficient, respectively, κ marks an elastic parameter as the function of ν and is taken as $1 - \nu$ for the plane strain state and $1/(1 + \nu)$ for the plane stress state.

One can From Eqs. (2.1) and (2.8) determine

$$\varepsilon_{\theta\theta} = \frac{1+\nu}{E\sqrt{2\pi r}}(K_I N_1 + K_{II} N_2 + T\sqrt{2\pi r}N_3) \tag{2.9}$$

in which

$$\begin{cases} N_I = \dfrac{1}{4}\left[(8\kappa - 5)\cos\dfrac{\theta}{2} + \cos\dfrac{3\theta}{2}\right] \\[2mm] N_3 = (\kappa - \cos^2\theta) \\[2mm] N_2 = -\dfrac{1}{4}\left[(8\kappa - 5)\sin\dfrac{\theta}{2} + 3\sin\dfrac{3\theta}{2}\right] \end{cases} \tag{2.10}$$

The initial fracturing propagation is imminent when the tangential strain $\varepsilon_{\theta\theta}$ along θ_c and at r_c reaches its maximum value $\varepsilon_{\theta\theta c}$, this implies that

$$\frac{\partial\varepsilon_{\theta\theta}}{\partial\theta}\Big|_{\theta=\theta_c} = 0 \text{ and } \frac{\partial^2\varepsilon_{\theta\theta}}{\partial^2\theta}\Big|_{\theta=\theta_c} < 0 \tag{2.11}$$

Introducing $\varepsilon_{\theta\theta}$ presented in Eq. (2.9) into Eq. (2.11) yields the angle for the fracture beginning in the form of normalized fracturing parameters as below:

$$K_I\left[(5 - 8\kappa)\sin\frac{\theta_c}{2} - 3\sin\frac{3\theta_c}{2}\right] + K_{II}\left[(5 - 8\kappa)\cos\frac{\theta_c}{2} - 9\cos\frac{3\theta_c}{2}\right]$$
$$+ 8T\sqrt{2\pi r_c}\sin 2\theta_c = 0 \tag{2.12}$$

For the pure mode-I fracturing problems (i.e., $K_I = K_{Ic}$, $K_{II} = 0$, and $\theta_c = 0$), Eq. (2.9) is implied as

$$E\varepsilon_{\theta\theta c}\sqrt{2\pi r_c} = (1 + \nu)\left[(2\kappa - 1) + \frac{T_0}{K_{Ic}}\sqrt{2\pi r_c}(\kappa - 1)\right]K_{Ic} \tag{2.13}$$

For the mixed-mode fracturing problems, we can conclude from Eqs. (2.9) and (2.13) that

$$
K_{Ic}(2\kappa - 1) + T_0\sqrt{2\pi r_c}(\kappa - 1) = K_I\frac{1}{4}\left[(8\kappa - 5)\cos\frac{\theta_c}{2} + \cos\frac{3\theta_c}{2}\right]
$$

$$
- K_{II}\frac{1}{4}\left[(8\kappa - 5)\sin\frac{\theta_c}{2} + 3\sin\frac{3\theta_c}{2}\right] + T\sqrt{2\pi r_c}(\kappa - \cos^2\theta_c)
$$

$$(2.14)$$

Analogously, the ratios of K_I to K_{Ic} and K_{II} to K_{Ic} are utilized to predict the fracture resistance of mixed-mode I/II as follows:

$$
\begin{cases}
\dfrac{K_I}{K_{Ic}} = \dfrac{(2\kappa - 1) + \frac{T_0}{K_{Ic}}\sqrt{2\pi r_c}(\kappa - 1)}{N_1(\theta_c) + \frac{K_{II}}{K_I}N_2(\theta_c) + \frac{T}{K_I}\sqrt{2\pi r_c}N_3(\theta_c)} \\[4mm]
\dfrac{K_{II}}{K_{Ic}} = \dfrac{(2\kappa - 1) + \frac{T_0}{K_{Ic}}\sqrt{2\pi r_c}(\kappa - 1)}{\frac{K_I}{K_{II}}N_1(\theta_c) + N_2(\theta_c) + \frac{T}{K_{II}}\sqrt{2\pi r_c}N_3(\theta_c)}
\end{cases}
$$

$$(2.15)$$

2.3 Maximum Tangential Strain Energy Density Factor (MTSEDF) Criterion

According to the definition of strain energy density (SED) [6, 7], one can derive

$$
\frac{dW}{dV} = \frac{1}{2}\sigma_{\theta\theta}\varepsilon_{\theta\theta}
$$

$$(2.16)$$

where dW/dV represents the density function of strain energy in the tangential direction. To represent the energy field intensity of the cracking tip vicinity, the tangential SED factor S_T is employed as

$$
S_T = r\frac{dW}{dV} = r\frac{1}{2}\sigma_{\theta\theta}\varepsilon_{\theta\theta}
$$

$$(2.17)$$

By introducing Eqs. (2.1) and (2.9) into Eq. (2.17), one can determine

$$
S_T = \frac{1}{8\pi G}[S_1 K_I^2 + S_2 K_{II}^2 + S_3 K_I K_{II} + S_4\sqrt{2\pi r}TK_I + S_5\sqrt{2\pi r}TK_{II} + S_6 2\pi r T^2]
$$

$$(2.18)$$

in which

$$
\begin{cases}
G = E/2(1+\nu) \\[4pt]
S_1 = \dfrac{1}{2}(\cos\dfrac{\theta}{2})^4(-3+4\kappa+\cos\theta) \\[8pt]
S_2 = \dfrac{3}{8}(\sin\theta)^2(-1+4\kappa+3\cos\theta) \\[8pt]
S_3 = -(\cos\dfrac{\theta}{2})^3(\sin\dfrac{\theta}{2})(-5+8\kappa+3\cos\theta) \\[8pt]
S_4 = -(\cos\dfrac{\theta}{2})^3[4-5\kappa+4(\kappa-1)\cos\theta+\cos2\theta] \\[8pt]
S_5 = (\cos\dfrac{\theta}{2})^2(\sin\dfrac{\theta}{2})[4-7\kappa+4(\kappa-1)\cos\theta+3\cos2\theta] \\[8pt]
S_6 = -\dfrac{1}{2}(\sin\theta)^2(1-2\kappa+\cos2\theta)
\end{cases}
\tag{2.19}
$$

The generalized MTSEDF criterion suggests that the initiation of fracturing growth is imminent when the value of S along θ_c and at r_c reaches its critical level S_c. This indicates that the sufficient conditions should be

$$
\frac{\partial S_T}{\partial\theta}\Big|_{\theta=\theta_c} = 0 \text{ and } \frac{\partial^2 S_T}{\partial^2\theta}\Big|_{\theta=\theta_c} < 0
\tag{2.20}
$$

By combining Eqs. (2.18) and (2.20), the fracturing deflection angle can be calculated from the following form:

$$
s_1 K_I^2 + s_2 K_{II}^2 + s_3 K_I K_{II} + s_4\sqrt{2\pi r_c}TK_I + s_5\sqrt{2\pi r_c}TK_{II} + s_6 2\pi r_c T^2 = 0
\tag{2.21}
$$

in which

$$
\begin{cases}
s_1 = (2\sin\theta_c+\sin2\theta_c)(5-8\kappa-3\cos\theta_c) \\[4pt]
s_2 = 3(\sin\theta_c)[3+4(4\kappa-1)\cos\theta_c+9\cos2\theta_c] \\[4pt]
s_3 = 4(\cos\dfrac{\theta_c}{2})^2[(13-16\kappa)(2\cos\theta_c-1)-9\cos2\theta_c] \\[8pt]
s_4 = 8(\cos\dfrac{\theta_c}{2})^2(\sin\dfrac{\theta_c}{2})[8-7\kappa+4(5\kappa-3)\cos\theta_c+7\cos2\theta_c] \\[8pt]
s_5 = 2(\cos\dfrac{\theta_c}{2})[(29-50\kappa)\cos\theta_c+(20\kappa-26)\cos2\theta_c+3(6\kappa-4+7\cos3\theta_c)] \\[8pt]
s_6 = 16(\sin2\theta_c)(\kappa-\cos2\theta_c)
\end{cases}
\tag{2.22}
$$

For the pure mode-I fracturing case, there are $K_{II}=0$, $K_I=K_{Ic}$, and $\theta_c=0$, then Eq. (2.21) can be reduced to

$$
S_{Tc} = \frac{S_0}{8\pi G}K_{Ic}^2
\tag{2.23}
$$

in which

$$S_0 = 2\kappa - 1 + (\kappa - 1)\frac{T_0}{K_{Ic}}\sqrt{2\pi r_c} \tag{2.24}$$

where T_0 symbolizes the T-stress value for the pure mode-I fracturing case. Under the combined-mode I/II loading state, substituting Eq. (2.23) into Eq. (2.18) yields

$$\frac{S_0}{8\pi G}K_{Ic}^2 = \frac{1}{8\pi G}[S_1 K_I^2 + S_2 K_{II}^2 +$$
$$S_3 K_I K_{II} + S_4\sqrt{2\pi r}TK_I + S_5\sqrt{2\pi r}TK_{II} + S_6 2\pi r T^2] \tag{2.25}$$

When both sides of Eq. (2.25) are divided by K_I^2, K_{II}^2 respectively, the fracture toughness ratios K_I/K_{Ic} and K_{II}/K_{Ic} can be obtained to forecast the combined-mode I/II fracturing onset as

$$
\begin{cases}
\dfrac{K_I}{K_{Ic}} = \left\{ S_0\left[S_1 + S_2\dfrac{K_{II}^2}{K_I^2} + S_3\dfrac{K_{II}}{K_I} + S_4\dfrac{T}{K_I}\sqrt{2\pi r_c} \right.\right. \\
\qquad\qquad \left.\left. +S_5\dfrac{K_{II}T}{K_I^2}\sqrt{2\pi r_c} + S_6\left(\dfrac{T}{K_I}\sqrt{2\pi r_c}\right)^2 \right] \right\}^{-1} \\[2em]
\dfrac{K_{II}}{K_{Ic}} = \left\{ S_0\left[S_1\dfrac{K_I^2}{K_{II}^2} + S_2 + S_3\dfrac{K_I}{K_{II}} + S_4\dfrac{K_I T}{K_{II}^2}\sqrt{2\pi r_c} \right.\right. \\
\qquad\qquad \left.\left. +S_5\dfrac{T}{K_{II}}\sqrt{2\pi r_c} + S_6\left(\dfrac{T}{K_{II}}\sqrt{2\pi r_c}\right)^2 \right] \right\}^{-1}
\end{cases} \tag{2.26}
$$

2.4 Minimum Strain Energy Density Factor (MSEDF) Criterion

According to the definition of strain energy density (SED) in an element [8], one can determine

$$\frac{dW}{dV} = \frac{1}{2G}\left[\frac{\eta + 1}{8}(\sigma_{rr} + \sigma_{\theta\theta})^2 - \sigma_{rr}\sigma_{\theta\theta} + \sigma_{r\theta}^2 \right] \tag{2.27}$$

where $\eta = (3 - 4v)$ for the plane strain case and $\eta = (3 - v)/(1 + v)$ for the plane stress case. To describe the energy field intensity at the neighbourhood of the cracking tip, the SED factor S is employed as

$$S = r\frac{dW}{dV} = \frac{r}{2G}\left[\frac{\eta+1}{8}(\sigma_{rr}+\sigma_{\theta\theta})^2 - \sigma_{rr}\sigma_{\theta\theta} + \sigma_{r\theta}^2\right] \qquad (2.28)$$

Introducing Eq. (2.1) into Eq. (2.28) yields

$$S = \frac{1}{16\pi G}[C_1 K_I^2 + C_2 K_I K_{II} + C_3 K_{II}^2$$
$$+ C_4 T\sqrt{2\pi r_c}K_I + C_5 T\sqrt{2\pi r_c}K_{II} + C_6(T\sqrt{2\pi r_c})^2] \qquad (2.29)$$

in which

$$\begin{cases} C_1 = (\eta - \cos\theta)(1 + \cos\theta) \\ C_2 = 2\sin\theta(2\cos\theta - \eta + 1) \\ C_3 = \eta(1 - \cos\theta) + \cos\theta(1 + 3\cos\theta) \\ C_4 = 2\cos\frac{\theta}{2}[\cos 2\theta - \cos\theta + \eta - 1] \\ C_5 = -2\sin\frac{\theta}{2}[\cos 2\theta + \cos\theta + \eta + 1] \\ C_6 = \frac{\eta+1}{2} \end{cases} \qquad (2.30)$$

The generalized MSEDF criterion demonstrates that the initiation of fracturing growth is imminent when the value of S along θ_c and at r_c reaches its critical level S_c. This implies that the sufficient conditions should be

$$\frac{\partial S}{\partial\theta}\bigg|_{\theta=\theta_c} = 0 \text{ and } \frac{\partial^2 S}{\partial\theta^2}\bigg|_{\theta=\theta_c} > 0 \qquad (2.31)$$

By combining Eqs. (2.29) and (2.31), the fracturing deflection angle can be calculated from the following form:

$$c_1 K_I^2 + c_2 K_I K_{II} + c_3 K_{II}^2 + c_4(T\sqrt{2\pi r_c})K_I + c_5(T\sqrt{2\pi r_c})K_{II} = 0 \qquad (2.32)$$

in which

$$\begin{cases} c_1 = \sin\theta_0(2\cos\theta_0 - \eta + 1) \\ c_2 = 4\cos(2\theta_0) + 2(1 - \eta)\cos\theta_0 \\ c_3 = -\sin\theta_0(6\cos\theta_0 - \eta + 1) \\ c_4 = \sin\frac{\theta_0}{2}(\frac{3}{2} - \eta) - \frac{5}{2}\sin(\frac{5\theta_0}{2}) \\ c_5 = -\cos\frac{\theta_0}{2}(\eta + \frac{1}{2}) - \frac{5}{2}\cos(\frac{5\theta_0}{2}) \end{cases} \qquad (2.33)$$

For the pure mode-I fracturing case, there are $K_{II} = 0$, $K_I = K_{Ic}$, and $\theta_c = 0$, then Eq. (2.29) can be reduced to

$$S_c = \frac{K_{Ic}^2}{16\pi G}C_0 \qquad (2.34)$$

in which

$$C_0 = 2(\eta - 1) + 2(\eta - 1)\frac{T_0}{K_{Ic}}\sqrt{2\pi r_c} + \frac{\eta + 1}{2}\left(\frac{T_0}{K_{Ic}}\sqrt{2\pi r_c}\right)^2 \tag{2.35}$$

Under the combined-mode I/II loading case, substituting Eq. (2.34) into Eq. (2.29) yields

$$C_1 K_I^2 + C_2 K_I K_{II} + C_3 K_{II}^2 + C_4\left(T\sqrt{2\pi r_c}\right)K_I$$
$$+ C_5\left(T\sqrt{2\pi r_c}\right)K_{II} + C_6\left(T\sqrt{2\pi r_c}\right)^2 = K_{Ic}^2 C_0 \tag{2.36}$$

When both sides of Eq. (2.36) are divided by K_I^2, K_{II}^2 respectively, the fracture toughness ratios K_I/K_{Ic} and K_{II}/K_{Ic} can be obtained to forecast the combined-mode I/II fracturing onset as

$$\begin{cases} \dfrac{K_I}{K_{Ic}} = \sqrt{\dfrac{C_0}{C_1 + C_2\frac{K_{II}}{K_I} + C_3\frac{K_{II}^2}{K_I^2} + C_4\frac{T\sqrt{2\pi r_c}}{K_I} + C_5\frac{T\sqrt{2\pi r_c}K_{II}}{K_I^2} + C_6\left(\frac{T\sqrt{2\pi r_c}}{K_I}\right)^2}} \\[4mm] \dfrac{K_{II}}{K_{Ic}} = \sqrt{\dfrac{C_0}{C_1\frac{K_I^2}{K_{II}^2} + C_2\frac{K_I}{K_{II}} + C_3 + C_4\frac{T\sqrt{2\pi r_c}K_I}{K_{II}^2} + C_5\frac{T\sqrt{2\pi r_c}}{K_{II}} + C_6\left(\frac{T\sqrt{2\pi r_c}}{K_{II}}\right)^2}} \end{cases} \tag{2.37}$$

2.5 Average Strain Energy Density (ASED) Criterion

The strain energy within the plastic or damaged region of radius r_c at the neighborhood of the crack tip is obtained from the integration method as follows [9]:

$$E(r_c) = \int_A \frac{dW}{dV}dA = \int_0^{r_c}\int_{-\pi}^{\pi}\frac{dW}{dV}rdrd\theta \tag{2.38}$$

Introducing Eq. (2.27) into Eq. (2.38) yields

$$E(r_c) = \frac{r_c}{16G}\Big[(2\eta - 1)K_I^2 + (2\eta + 3)K_{II}^2$$
$$+ \frac{16}{15\pi}(5\eta - 7)K_I T\sqrt{2\pi r} + \frac{(\eta + 1)}{2}\left(T\sqrt{2\pi r}\right)^2\Big] \tag{2.39}$$

Therefore, the average strain energy density (ASED) on the plastic or damaged zone with the radius r_c can be defined as below:

$$\overline{E} = \frac{E(r_c)}{\pi r_c^2} = \frac{1}{16G\pi r_c(2\eta - 1)} \left[K_I^2 + \frac{(2\eta + 3)}{(2\eta - 1)} K_{II}^2 \right.$$

$$\left. + \frac{16(5\eta - 7)}{15\pi(2\eta - 1)} T\sqrt{2\pi r_c}K_I + \frac{(\eta + 1)}{2(2\eta - 1)}(T\sqrt{2\pi r_c})^2 \right] \qquad (2.40)$$

For the pure mode-I fracturing case, there are $K_{II} = 0$, $K_I = K_{Ic}$, and $\theta_c = 0$, then Eq. (2.40) can be reduced to

$$\overline{E}_c = \frac{K_{Ic}^2}{16\mu\pi r_0(2\eta - 1)} A_0$$

$$= \frac{K_{Ic}^2}{16\mu\pi r_0(2\eta - 1)} \left[1 + \frac{16(5\eta - 7)}{15\pi(2\eta - 1)} \frac{T_0}{K_{Ic}} \sqrt{2\pi r_c} + \frac{(\eta + 1)}{2(2\eta - 1)}(\frac{T_0}{K_{Ic}}\sqrt{2\pi r_c})^2 \right]$$

$$(2.41)$$

Introducing Eq. (2.41) into Eq. (2.40) yields

$$K_I^2 + \frac{(2\eta + 3)}{(2\eta - 1)} K_{II}^2 + \frac{16(5\eta - 7)}{15\pi(2\eta - 1)} T\sqrt{2\pi r_c}K_I + \frac{(\eta + 1)}{2(2\eta - 1)}(T\sqrt{2\pi r_c})^2 = K_{Ic}^2 A_0$$

$$(2.42)$$

When both sides of Eq. (2.42) are divided by K_I^2, K_{II}^2 respectively, the fracture toughness ratios K_I/K_{Ic} and K_{II}/K_{Ic} can be acquired to evaluate the onset of combined-mode I/II fracturing as

$$\begin{cases} \dfrac{K_I}{K_{Ic}} = \sqrt{\dfrac{A_0}{1 + \frac{(2\eta+3)}{(2\eta-1)} \frac{K_{II}^2}{K_I^2} + \frac{16(5\eta-7)}{15\pi(2\eta-1)} \frac{T\sqrt{2\pi r_c}}{K_I} + \frac{(\eta+1)}{2(2\eta-1)} \left(\frac{T\sqrt{2\pi r_c}}{K_I}\right)^2}} \\[4ex] \dfrac{K_{II}}{K_{Ic}} = \sqrt{\dfrac{A_0}{\frac{K_I^2}{K_{II}^2} + \frac{(2\eta+3)}{(2\eta-1)} + \frac{16(5\eta-7)}{15\pi(2\eta-1)} \frac{K_I T\sqrt{2\pi r_c}}{K_{II}^2} + \frac{(\eta+1)}{2(2\eta-1)} \left(\frac{T\sqrt{2\pi r_c}}{K_{II}}\right)^2}} \end{cases} \qquad (2.43)$$

2.5.1 Maximum Energy Release Rate (MERR) Criterion

In compliance with three significant factors: the definition of SIF (stress intensity factor), a continuity hypothesis, and the correlation between fracture toughness and ERR (energy release rate), the ERR for the branching crack (see Fig. 2.2 [10]) can be determined from

$$G_\theta = \frac{2\pi r}{E'}(\sigma_{\theta\theta}^2 + \sigma_{r\theta}^2) \qquad (2.44)$$

Fig. 2.2 Geometry and polar coordinate system for a branching crack [10]

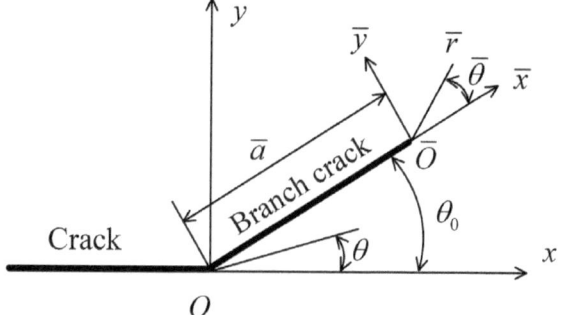

where $E' = E$ for the plane stress case and $E' = E/(1-2v)$ for the plane strain case. Introducing Eq. (1) into Eq. (2.44) yields

$$G_\theta = \frac{1}{E'}[G_1 K_I^2 + G_2 K_I K_{II} + G_3 K_{II}^2$$
$$+ G_4 T\sqrt{2\pi r_c}K_I + G_5 T\sqrt{2\pi r_c}K_{II} + G_6(T\sqrt{2\pi r_c})^2] \qquad (2.45)$$

in which

$$\begin{cases} G_1 = \frac{1}{4}(1 + \cos\theta)^2 \\ G_2 = -\frac{1}{2}\sin(2\theta) - \sin\theta \\ G_3 = -3\sin^4\frac{\theta}{2} + 2\sin^2\frac{\theta}{2} + 1 \\ G_4 = -4\cos^5\frac{\theta}{2} + 4\cos^3\frac{\theta}{2} \\ G_5 = 4\sin^5\frac{\theta}{2} - 4\sin\frac{\theta}{2} \\ G_6 = \sin^2\theta \end{cases} \qquad (2.46)$$

The generalized MERR criterion demonstrates that the initiation of fracturing growth is imminent when the value of S along θ_c and at r_c approaches its critical level $G_{\theta c}$. This implies that the sufficient conditions should be

$$\frac{\partial G_\theta}{\partial\theta}\Big|_{\theta=\theta_c} = 0 \text{ and } \frac{\partial^2 G_\theta}{\partial\theta^2}\Big|_{\theta=\theta_c} > 0 \qquad (2.47)$$

By combining Eqs. (2.45) and (2.47), the crack deflection angle θ_c can be calculated from the following form:

$$g_1 K_I^2 + g_2 K_I K_{II} + g_3 K_{II}^2 + g_4 T \sqrt{2\pi r_c} K_I + g_5 T \sqrt{2\pi r_c} K_{II} + g_6 T \sqrt{2\pi r_c}^2 = 0 \tag{2.48}$$

in which

$$\begin{cases} g_1 = -\frac{1}{4} \sin 2\theta - \frac{1}{2} \sin \theta \\ g_2 = -\cos(2\theta) - \cos \theta \\ g_3 = \frac{3}{4} \sin 2\theta - \frac{1}{2} \sin \theta \\ g_4 = 10 \cos^4 \frac{\theta}{2} \sin \frac{\theta}{2} - 6 \cos^2 \frac{\theta}{2} \sin \frac{\theta}{2} \\ g_5 = 10 \sin^4 \frac{\theta}{2} \cos \frac{\theta}{2} - 2\cos \frac{\theta}{2} \\ g_6 = \sin 2\theta \end{cases} \tag{2.49}$$

For the pure mode-I fracturing case, there are $K_{II} = 0$, $K_I = K_{Ic}$, and $\theta_c = 0$, then Eq. (2.45) can be simplied as

$$G_{\theta c} = \frac{K_{Ic}^2}{E'} \tag{2.50}$$

Introducing Eq. (2.50) into Eq. (2.45) yields

$$G_1 K_I^2 + G_2 K_I K_{II} + G_3 K_{II}^2 + G_4 T \sqrt{2\pi r_c} K_I + G_5 T \sqrt{2\pi r_c} K_{II} + G_6 (T \sqrt{2\pi r_c})^2 = K_{Ic}^2 \tag{2.51}$$

When both sides of Eq. (2.51) are divided by K_I^2, K_{II}^2 respectively, the fracture toughness ratios K_I/K_{Ic} and K_{II}/K_{Ic} can be acquired to estimate the onset of combined-mode I/II fracturing as

$$\begin{cases} \frac{K_I}{K_{Ic}} = \sqrt{\dfrac{1}{G_1 + G_2 \frac{K_{II}}{K_{II}} + G_3 \frac{K_{II}^2}{K_I^2} + G_4 \frac{T\sqrt{2\pi r_c}}{K_I} + G_5 \frac{T\sqrt{2\pi r_c}K_{II}}{K_I^2} + G_6 \left(\frac{T\sqrt{2\pi r_c}}{K_I}\right)^2}} \\ \frac{K_{II}}{K_{Ic}} = \sqrt{\dfrac{1}{G_1 \frac{K_I^2}{K_{II}^2} + G_2 \frac{K_I}{K_{II}} + G_3 + G_4 \frac{T\sqrt{2\pi r_c}K_I}{K_{II}^2} + G_5 \frac{T\sqrt{2\pi r_c}}{K_{II}} + G_6 \left(\frac{T\sqrt{2\pi r_c}}{K_{II}}\right)^2}} \end{cases} \tag{2.52}$$

Traditionally, the critical microcracking length r_c is proverbially recognized to be the magnitude of the fracture process zone (FPZ) for rocks, which can be characterized by associating K_{Ic} with σ_t considering the maximum tension stress law as

$$r_c = \frac{1}{2\pi} \left(\frac{K_{Ic}}{\sigma_t}\right)^2 \tag{2.53}$$

where $\sigma_t = P_{max}/\pi BR$ symbolizes the tensile strength evaluated indirectly from the unnotched Brazilian disk testing.

As documented previously, the stress-based fracture criteria are not more complicated compared to the energy-based ones. Nevertheless, the energy-based crireria exhibit a greater accuracy in predicting fracture, the strain-based criteria are characterized by the combinations of both the abovementioned models.

2.6 3D Fracture Criteria

Under the combined-mode I/III loading, the singular terms for modes I and III certainly are a predominant factor representing the stress field at the vicinity of a cracking tip. Truncating the singular terms in the Williams' solution yields the stress field in a cylindrical coordinate system (see Fig. 2.3 [11]) as follows:

$$
\begin{cases}
\sigma_{rr} = \dfrac{K_I}{\sqrt{2\pi r}} \cos\dfrac{\theta}{2}\left(1 + \sin^2\dfrac{\theta}{2}\right) \\[2mm]
\sigma_{\theta\theta} = \dfrac{K_I}{\sqrt{2\pi r}} \cos^3\dfrac{\theta}{2} \\[2mm]
\tau_{r\theta} = \dfrac{K_I}{2\sqrt{2\pi r}} \cos\dfrac{\theta}{2}\sin\theta \\[2mm]
\tau_{rz} = \dfrac{K_{III}}{\sqrt{2\pi r}} \sin\dfrac{\theta}{2} \\[2mm]
\tau_{\theta z} = \dfrac{K_{III}}{\sqrt{2\pi r}} \cos\dfrac{\theta}{2} \\[2mm]
\sigma_{zz} = 0
\end{cases}
\tag{2.54}
$$

where K_I and K_{III} denote respectively the modes I and III SIFs. Note that the expression $\sigma_{zz} = 0$ for the plane stress state can be substituted by $\sigma_{zz} = \nu(\sigma_{rr} + \sigma_{\theta\theta})$ for the plane strain state.

To make the fracture criteria competent in predicting the fracture resistance and crack initiation angle for combined-mode I/III fracture problems, supposing that the initial fracture state at the θ-z plane is rotated by a specific angle φ around

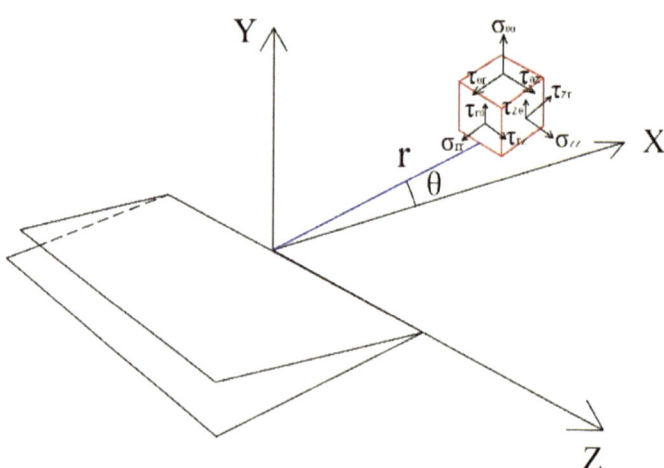

Fig. 2.3 The stress field acting on an element at the vicinity of a cracking tip

Fig. 2.4 Stress components at the θ−z plane

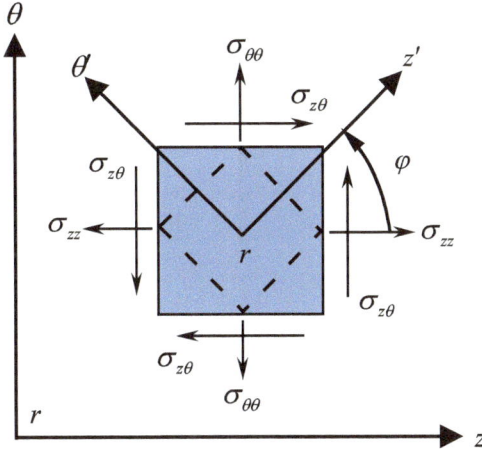

the r-coordinate axis to obtain the equivalent element, as described in Fig. 2.4. In conjunction with Eq. (2.54), an original stress tensor σ can be determined as

$$\sigma = \begin{bmatrix} \sigma_{rr} & \tau_{r\theta} & \tau_{rz} \\ \tau_{r\theta} & \sigma_{\theta\theta} & \tau_{\theta z} \\ \tau_{rz} & \tau_{\theta z} & \sigma_{zz} \end{bmatrix} \tag{2.55}$$

Applying the transformation relation between the (r, θ, z) and (r', θ', z') coordinates yields another stress tensor σ':

$$\sigma' = \Omega \sigma \Omega^T = \begin{bmatrix} \sigma_{rr} & \tau_{r\theta'} & \tau_{rz'} \\ \tau_{r\theta'} & \sigma_{\theta'\theta'} & \tau_{\theta'z'} \\ \tau_{rz'} & \tau_{\theta'z'} & \sigma_{z'z'} \end{bmatrix} \tag{2.56}$$

where Ω symbolizes the transition matrix, which is written by

$$\Omega = \begin{bmatrix} 1 & 0 & 0 \\ 0 & \cos\varphi & -\sin\varphi \\ 0 & \sin\varphi & \cos\varphi \end{bmatrix} \tag{2.57}$$

Hence, the normal stresses $\sigma_{\theta'\theta'}$ and $\sigma_{z'z'}$ can be respectively determined from Eqs. (2.54)–(2.57) as follows [12, 13]:

$$\sigma_{\theta'\theta'} = \frac{1}{\sqrt{2\pi r}} (K_I \cos^2\varphi \cos^3\frac{\theta}{2} - K_{III} \sin 2\varphi \cos\frac{\theta}{2}) \tag{2.58}$$

$$\sigma_{z'z'} = \frac{1}{\sqrt{2\pi r}} K_I \sin^2\varphi \cos^3\frac{\theta}{2} + \frac{K_{III}}{\sqrt{2\pi r}} \sin 2\varphi \cos\frac{\theta}{2} \tag{2.59}$$

Based on the elastic model of Hooke [14, 15], the tangential strain $\varepsilon_{\theta'\theta'}$ is expressed as follows:

$$\varepsilon_{\theta'\theta'} = \frac{1}{E}[\sigma_{\theta'\theta'} - \nu(\sigma_{rr} + \sigma_{z'z'})] \tag{2.60}$$

where E and ν stand for respectively the Young and Poisson coefficients, introducing Eqs. (2.54), (2.58), and (2.59) into Eq. (2.60) yields

$$\varepsilon_{\theta'\theta'} = \frac{1}{E\sqrt{2\pi r}}\left\{K_I[(1 + \nu)\cos^2\varphi\cos^3\frac{\theta}{2} - 2\nu\cos\frac{\theta}{2}] - K_{III}(1 + \nu)\sin 2\varphi\cos\frac{\theta}{2}\right\} \tag{2.61}$$

According to the definition of SED in the tangential direction, one can determine

$$\frac{dW}{dV} = \frac{1}{2}\sigma_{\theta'\theta'}\varepsilon_{\theta'\theta'} \tag{2.62}$$

where dW/dV represents the function of SED. With reference to the mathematical model first proposed by Sih [16], to represent the energy field intensity of the notch tip vicinity, the tangential SED factor C_T is adopted as follows [17, 18]:

$$C_T = r\frac{dW}{dV} = r\frac{1}{2}\sigma_{\theta\theta}\varepsilon_{\theta\theta} \tag{2.63}$$

By introducing Eqs. (2.54) and (2.61) into Eq. (2.63), one can determine

$$C_T = \frac{1}{4\pi E}(C_{T1}K_I^2 + C_{T2}K_I K_{III} + C_{T3}K_{III}^2) \tag{2.64}$$

in which

$$\begin{cases} C_{T1} = \cos^2\varphi\cos^4\frac{\theta}{2}[(1 + \nu)\cos^2\varphi\cos^2\frac{\theta}{2} - 2\nu] \\ C_{T2} = -2\sin 2\varphi\cos^2\frac{\theta}{2}[(1 + \nu)\cos^2\varphi\cos^2\frac{\theta}{2} - \nu] \\ C_{T3} = (1 + \nu)\sin^2 2\varphi\cos^2\frac{\theta}{2} \end{cases} \tag{2.65}$$

The 3D-MTS, 3D-MTSN, and 3D-MTSEDF criteria suggest that the beginning of fracturing extension is imminent when the values of $\sigma_{\theta'\theta'}$, $\varepsilon_{\theta'\theta'}$, and C_T along initial cracking direction and at critical damaged radius exceed the critical level. Using the extremum theorem, the planar fracture deflection angle θ_c and the nonplanar fracture twist angle φ_c are solved by differentiating Eqs. (2.54), (2.60), and (2.63) relative to φ and θ as below:

$$\theta_c = 0 \tag{2.66}$$

$$\varphi_c = \frac{1}{2} arc \tan \frac{-2K_{III}}{K_I} \tag{2.67}$$

Under the pure mode-I loading, there are $K_{III} = \theta_c = \varphi_c = 0$ and $K_I = K_{Ic}$, then the three critical values can be computed as below:

$$\begin{cases} \sigma_{\theta\theta Ic} = \frac{1}{\sqrt{2\pi r_c}} K_{Ic} \\ \tau_{\theta\theta Ic} = \frac{1-\nu}{E\sqrt{2\pi r_c}} K_{Ic} \\ C_{Tc} = \frac{1-\nu}{4\pi E} K_{Ic}^2 \end{cases} \tag{2.68}$$

Under the pure mode-III loading, we can conclude that $\varphi_c = -45°$, $K_I = 0$, and $K_{III} = K_{IIIc}$. Particularly, the predicted values of K_{IIIc}/K_{Ic} can be determined from the three leading fracture criteria as follows:

$$\begin{cases} \dfrac{K_{IIIc}^{3D-MTS}}{K_{Ic}} = 1 \\ \dfrac{K_{IIIc}^{3D-MTSN}}{K_{Ic}} = \dfrac{1-\nu}{1+\nu} \\ \dfrac{K_{IIIc}^{3D\text{-}MTSEDF}}{K_{Ic}} = \sqrt{\dfrac{1-\nu}{1+\nu}} \end{cases} \tag{2.69}$$

Consequently, the envelopes for theoretical fracture resistance can be depicted by the fracture resistance ratios K_I/K_{Ic} and K_{III}/K_{Ic} extracted from the above-mentioned formulae as below:

3D-MTS criterion:

$$\begin{cases} \dfrac{K_I}{K_{Ic}} = (\cos^2 \varphi_c - \dfrac{K_{III}}{K_I} \sin 2\varphi_c)^{-1} \\ \dfrac{K_{III}}{K_{Ic}} = (\dfrac{K_I}{K_{III}} \cos^2 \varphi_c - \sin 2\varphi_c)^{-1} \end{cases} \tag{2.70}$$

3D-MTSN criterion:

$$\begin{cases} \dfrac{K_I}{K_{Ic}} = (\dfrac{1+\nu}{1-\nu} \cos^2 \varphi_c - \dfrac{K_{III}}{K_I}\dfrac{1+\nu}{1-\nu} \sin 2\varphi_c - 2\dfrac{\nu}{1-\nu})^{-1} \\ \dfrac{K_{III}}{K_{Ic}} = (\dfrac{K_I}{K_{III}}\dfrac{1+\nu}{1-\nu} \cos^2 \varphi_c - \dfrac{1+\nu}{1-\nu} \sin 2\varphi_c - 2\dfrac{K_I}{K_{III}}\dfrac{\nu}{1-\nu})^{-1} \end{cases} \tag{2.71}$$

3D-MTSEDF criterion:

$$\begin{cases} \dfrac{K_I}{K_{Ic}} = \\[4pt] \left(\dfrac{1-\nu}{\cos^2\varphi_c[(1+\nu)\cos^2\varphi_c - 2\nu] - 2\sin 2\varphi_c[(1+\nu)\cos^2\varphi_c - \nu]\frac{K_{III}}{K_I} + (1+\nu)\sin^2 2\varphi_c(\frac{K_{III}}{K_I})^2} \right)^{\frac{1}{2}} \\[10pt] \dfrac{K_{III}}{K_{Ic}} = \\[4pt] \left(\dfrac{1-\nu}{\cos^2\varphi_c[(1+\nu)\cos^2\varphi_c - 2\nu](\frac{K_L}{K_{III}})^2 - 2\sin 2\varphi_c[(1+\nu)\cos^2\varphi_c - \nu]\frac{K_L}{K_{III}} + (1+\nu)\sin^2 2\varphi_c} \right)^{\frac{1}{2}} \end{cases}$$

$$(2.72)$$

References

1. Williams ML (1957) On the stress distribution at the base of a stationary crack. J Appl Mech 24:109–114
2. Kun Z, Zhao Yu, Chaolin W, Jing Bi (2024) Influence of distinct testing methods on the mode-I fracture toughness of Longmaxi shale. Theoret Appl Fract Mech 129:104213
3. Smith DJ, Ayatollahhi MR, Pavier MJ (2001) The role of T-stress in brittle fracture for linear elastic materials under mixed-mode loading, Fatigue. Fract Engng Mater Struct 24:137–150
4. Mirsayar MM (2015) Mixed mode fracture analysis using extended maximum tangential strain criterion. Mater Design 86:941–947
5. Hua W, Dong S, Pan X, Wang Q (2017) Mixed mode fracture analysis of CCBD specimens based on the extended maximum tangential strain criterion. Fatigue Fract Engng Mater Struct 00:1–10
6. Ayatollahi MR, Saboori B (2014) Maximum tangential strain energy density criterion for general mixed mode I/II/III brittle fracture. Int J Damage Mech 1–16
7. Liu J, Qiao L, Li Y, Li QW, Fan DJ (2022) Experimental study on the quasi-static loading rate dependency of mixed-mode I/II fractures for marble rocks. Theor Appl Fract Mec 121:103431
8. Ayatollahi MR, Rashidi Moghaddam M, Berto F (2015) A generalized strain energy density criterion for mixed mode fracture analysis in brittle and quasi-brittle materials. Theor Appl Fract Mech 79:70–76
9. Rashidi Moghaddam M, Ayatollahi MR, Berto F (2017) Mixed mode fracture analysis using generalized averaged strain energy density criterion for linear elastic materials. Int J Solids Struct 120:137–145
10. Hou C, Jin XC, Fan XL, Xu R, Wang ZY (2019) A generalized maximum energy release rate criterion for mixed mode fracture analysis of brittle and quasi-brittle materials. Theor Appl Fract Mec 100:78–85
11. Zheng K, Chaolin Wang Yu, Zhao JB, Liu H (2024) A modified three-dimensional mean strain energy density criterion for predicting shale mixed-mode I/III fracture toughness. J Rock Mech Geotech Eng. https://doi.org/10.1016/j.jrmge.2023.09.016
12. Liu S, Chao YJ, Zhu X (2004) Tensile-shear transition in mixed mode I/III fracture. Int J Solids Struct 41(2004):6147–6172
13. Ayatollahi MR, Saboori B (2015) T-stress effects in mixed mode I/II/III brittle fracture. Eng Fract Mech 144:32–45
14. Bidadi J, Aliha MRM, Akbardoost J (2022) Development of maximum tangential strain (MTSN) criterion for prediction of mixed-mode I/III brittle fracture. Int J Solids Struct 256:111979
15. Tang W, Zhang Y, Zhao Y, Zheng K, Wang C, Bi J (2024) Assessment of basalt fiber and gelling enhancement effects on mixed mode I/III fracture performance of the mortar composites 104303

16. Sih GC (1974) Strain-energy-density factor applied to mixed mode crack problems. Int J Fract 10:305–321
17. Ayatollahi M, Saboori B (2015) Maximum tangential strain energy density criterion for general mixed mode I/II/III brittle fracture. Int J Damage Mech 24:263–278
18. Hua W, Huang JZ, Pan X, Li JX, Dong SM (2021) An extended maximum tangential strain energy density criterion considering T-stress for combined mode I-III brittle fracture. Fatig Fract Eng Mater Struct 44:169–181

Chapter 3
Review of Fracture Test Methods

3.1 Test Methods of Mixed-Mode I/II Fracture

The first test specimen, known as the CCBD (centrally cracked Brazilian disk) specimen, is widely employed for evaluating the mixed-mode I/II fracture resistance of different materials. This specimen is a disk of diameter D and thickness B with a straight crack of length $2a$ created at the specimen's center. By applying diametrical compression loads, the complete range of mixed-mode I/II fracture can be accomplished depending on the loading angle β, which is also regarded as the crack orientation angle between the prefabricated notch direction and the loading line. The mixed-mode I/II fracture parameters for the CCBD specimen are determined from [1, 2]:

$$
\begin{cases}
K_I = \dfrac{P}{\pi BR}\sqrt{\pi a}F_I \\[2mm]
K_{II} = \dfrac{P}{\pi BR}\sqrt{\pi a}F_{II} \\[2mm]
K_e = \sqrt{K_I^2 + K_{II}^2} \\[2mm]
M^e = \dfrac{2}{\pi}\tan^{-1}(\dfrac{K_I}{K_{II}}) = \dfrac{2}{\pi}\tan^{-1}(\dfrac{F_I}{F_{II}}) \\[2mm]
T = \dfrac{P}{\pi BR}T^*
\end{cases}
\tag{3.1}
$$

where K_I and K_{II} mean respectively the modes I and II SIFs, P symbolizes the applied compression load, B and R mark respectively the thickness and radius of the CCBD specimen, F_I and F_{II} denote respectively the non-dimensional geometric factors for modes I and II, a stands for the prefabricated crack length, T^* is the non-dimensional expression of the T-stress, K_e reprensents the effective SIF, and M^e symbolizes the mode mixity index, which is regarded as 0 for pure mode-II loading case and 1 for pure mode-I loading case.

© The Author(s) 2024
Y. Zhao et al., *Rock Fracture Mechanics and Fracture Criteria*,
https://doi.org/10.1007/978-981-97-5822-7_3

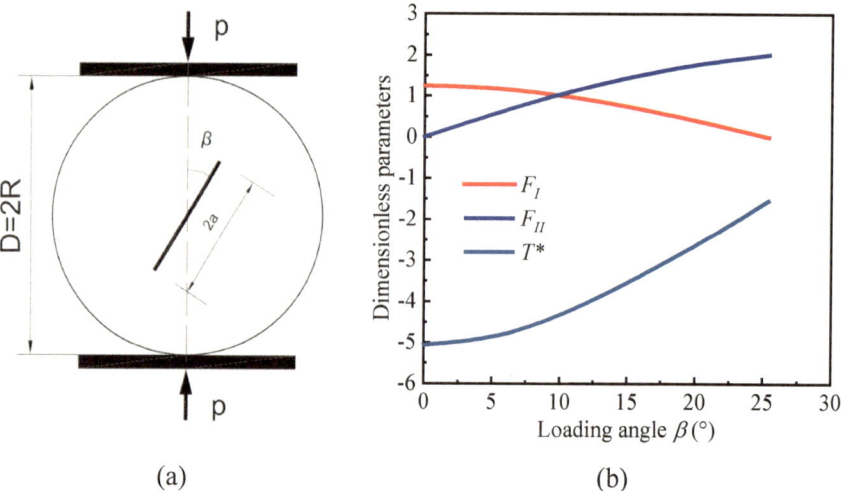

(a)   (b)

Fig. 3.1 Schematic of the CCBD specimen and dimensionless fracture parameters

The schematic of the CCBD specimen is displayed in Fig. 3.1a. Taking the CCBD specimen with $a/R = 0.4$ as an example, the variations of F_I, F_{II}, and T^* versus the loading angle β are outlined in Fig. 3.1b [3, 4]. Particularly, the loading angle $\beta = 0°$ corresponds to the pure mode-I loading case, while the loading angle $\beta = 25.4°$ corresponds to the pure mode-II loading case.

The second test specimen, known as the SCB (semi-circular bending) specimen, is a semi-disk of diameter D and thickness B and contains a straight edge notch of length a processed at the specimen's bottom. When conducting this testing procedure, the SCB sample is supported by two symmetrical rollers with support span S, and the middle roller is employed to produce the compressive force. Consequently, then the whole range of mixed-mode I/II fracture can be accomplished depending on the loading angle, which is also regarded as the crack orientation angle between the prefabricated notch direction and the loading line. The mixed-mode I/II fracture parameters for the SCB specimen are determined from [5, 6]:

$$
\begin{cases}
K_I = \dfrac{P}{2RB}\sqrt{\pi a}F_I \\[2mm]
K_{II} = \dfrac{P}{2RB}\sqrt{\pi a}F_{II} \\[2mm]
K_E = \sqrt{K_I^2 + K_{II}^2} \\[2mm]
M^e = \dfrac{2}{\pi}\tan^{-1}(\dfrac{K_I}{K_{II}}) = \dfrac{2}{\pi}\tan^{-1}(\dfrac{F_I}{F_{II}}) \\[2mm]
T = \dfrac{P}{2RB}T^*
\end{cases}
\tag{3.2}
$$

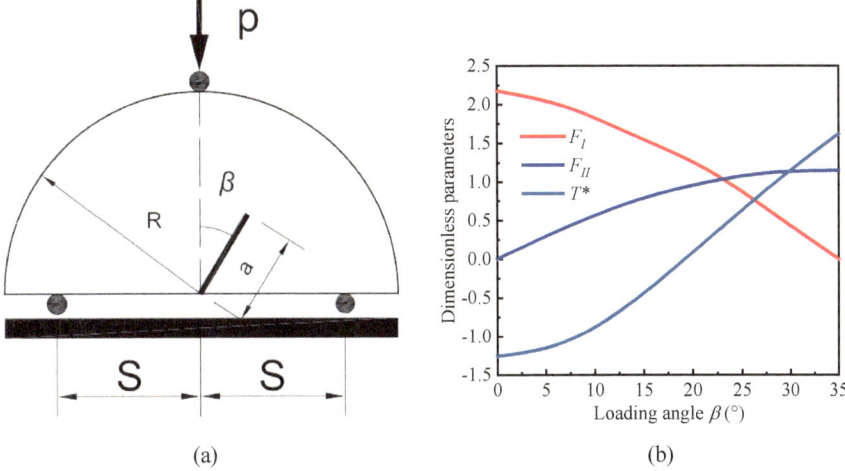

(a)   (b)

Fig. 3.2 Schematic of the SCB specimen and dimensionless fracture parameters

The schematic of the SCB specimen is displayed in Fig. 3.2a. Taking the SCB specimen with $a/R = 0.4$ and $S/R = 0.4$ as an example, the variations of F_I, F_{II}, and T^* versus the loading angle β are outlined in Fig. 3.2b [7, 8]. Particularly, the loading angle $\beta = 0°$ corresponds to the pure mode-I loading case, while the loading angle $\beta = 35°$ corresponds to the pure mode-II loading case.

The third test specimen, known as the SBB (short beam bending) specimen, is a rectangular beam of height W and thickness B and contains a straight edge notch of length a processed at the specimen's bottom. When conducting this testing procedure, the SBB sample is supported by two symmetrical rollers with support span $2S$, and the middle roller is employed to produce the compressive force. Consequently, then the entire range of mixed-mode I/II fracture can be accomplished depending on the loading angle, which is also regarded as the crack orientation angle between the prefabricated notch direction and the loading line. The mixed-mode I/II fracture parameters for the SBB specimen are determined from [9, 10]:

$$
\begin{cases}
K_I = \dfrac{P}{WB}\sqrt{\pi a}F_I \\[2mm]
K_{II} = \dfrac{P}{WB}\sqrt{\pi a}F_{II} \\[2mm]
K_e = \sqrt{K_I^2 + K_{II}^2} \\[2mm]
M^e = \dfrac{2}{\pi}\tan^{-1}(\dfrac{K_I}{K_{II}}) = \dfrac{2}{\pi}\tan^{-1}(\dfrac{F_I}{F_{II}}) \\[2mm]
T = \dfrac{P}{WB}T^*
\end{cases}
\tag{3.3}
$$

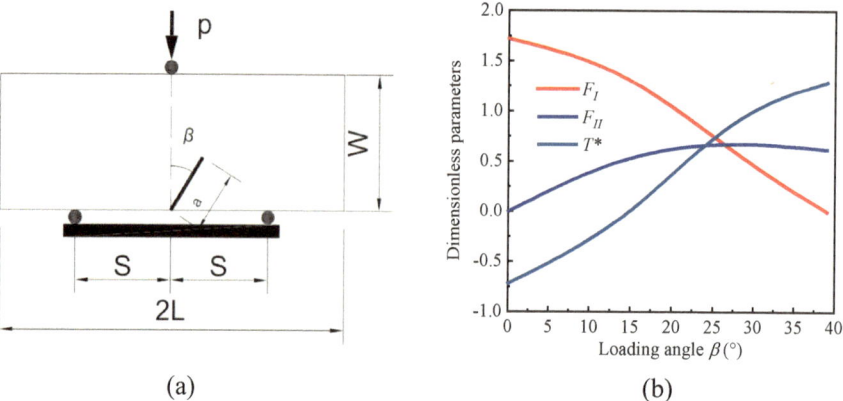

Fig. 3.3 Schematic of the SBB specimen and dimensionless fracture parameters

The schematic of the SBB specimen is displayed in Fig. 3.3a. Taking the SBB specimen with $a/W = 0.5$ and $S/W = 0.5$ as an example, the variations of F_I, F_{II}, and T^* versus the loading angle β are outlined in Fig. 3.3b [11, 12]. Particularly, the loading angle $\beta = 0°$ corresponds to the pure mode-I loading case, while the loading angle $\beta = 39°$ corresponds to the pure mode-II loading case.

The forth test specimen, known as the ECTB (edge-cracked triangular bending) specimen, is a triangular plate of thickness B and base length $2W$ and contains a straight edge notch of length a processed at the specimen's bottom. When conducting this testing procedure, the ECTB sample is supported by two symmetrical rollers with support span $2S$, and the middle roller is employed to produce the compressive force. Consequently, then the full range of mixed-mode I/II fracture can be accomplished depending on the loading angle, which is also regarded as the crack orientation angle between the prefabricated notch direction and the loading line. The mixed-mode I/II fracture parameters for the ECTB specimen are determined from [13, 14]:

$$\begin{cases} K_I = \dfrac{P}{2WB}\sqrt{\pi a}F_I \\[2mm] K_{II} = \dfrac{P}{2WB}\sqrt{\pi a}F_{II} \\[2mm] K_e = \sqrt{K_I^2 + K_{II}^2} \\[2mm] M^e = \dfrac{2}{\pi}\tan^{-1}(\dfrac{K_I}{K_{II}}) = \dfrac{2}{\pi}\tan^{-1}(\dfrac{F_I}{F_{II}}) \\[2mm] T = \dfrac{P}{2WB}T^* \end{cases} \qquad (3.4)$$

The schematic of the ECTB specimen is displayed in Fig. 3.4a. Taking the SBB specimen with $a/W = 0.3$ and $S/W = 0.4$ as an example, the variations of F_I, F_{II}, and T^* versus the loading angle β are outlined in Fig. 3.4b [15]. Particularly, the

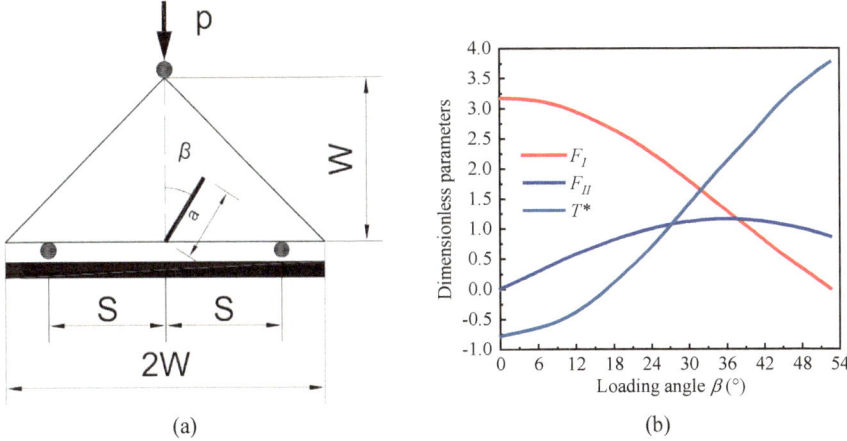

(a)   (b)

Fig. 3.4 Schematic of the ECTB specimen and dimensionless fracture parameters

loading angle $\beta = 0°$ corresponds to the pure mode-I loading case, while the loading angle $\beta = 52.5°$ corresponds to the pure mode-II loading case.

3.2 Test Methods of True Mode-II Fracture

The first test specimen, known as the SCC (short core in compression) specimen, is a cylinder of height H and diameter D, and contains two horizontal half-through notches of depth a from opposite sides and a ligament of vertical length C between the two notches [16–22]. By applying the uniaxial compression loads, the rectangular rock bridge in the axial plane of the SCC specimen can be broken, so the shear-induced fracture trajectory is self-planar and self-similar.

The schematic of the SCC specimen is displayed in Fig. 3.5a, the variation of F_{II} versus the C/H values is outlined in Fig. 3.5b. According to the recommendation of Xu et al. [16], the SCC specimen with $H/D = 2$, $C/H = 0.2$, and notch thickness $t = 1$ mm is relatively reliable and optimal for the measurement of true mode-II fracture resistance. The true mode-II fracture toughness for the SCC specimen is determined from [16]:

$$K_{IIc} = \frac{P}{CD\sqrt{\pi a}}F_{II} = \frac{P}{CD\sqrt{\pi a}}(2.3744\frac{C}{H} + 0.0192) \tag{3.5}$$

The second test specimen, known as the SB (shear-box) specimen, is a cube of width W thickness B, and length L, and contains the single-edge notch of inclined depth a or the double-edge notch of inclined depth $2a$ and a ligament [23–28]. By

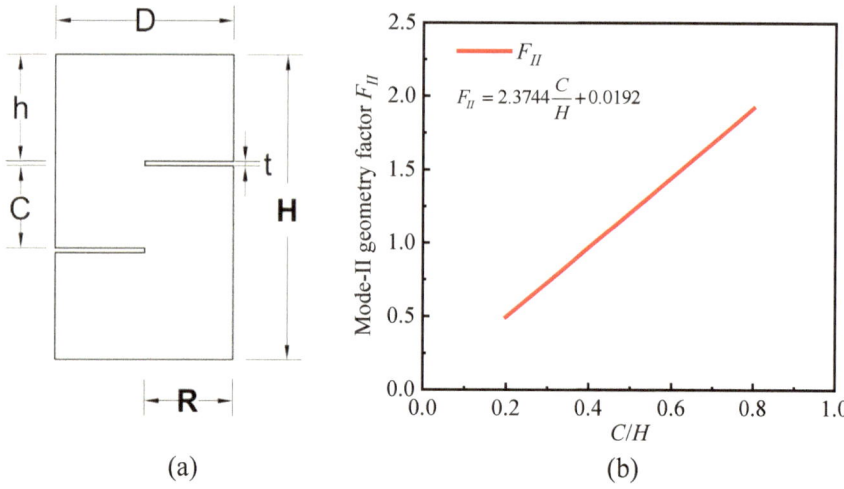

Fig. 3.5 Schematic of the SCC specimen and dimensionless fracture parameters

applying the uniaxial compression loads, the rectangular rock bridge in the middle plane of the SB specimen can be broken, so the shear-induced fracture trajectory is self-planar and self-similar.

The schematic of the SB specimen is displayed in Fig. 3.6. According to the recommendation of Rao et al. [23], the single-edge notched SB specimen with $L = W = B$, $a/W \geq 0.5$, and notch thickness $t = 1$ mm is relatively reliable and optimal for the measurement of true mode-II fracture resistance. While the double-edge notched SB specimen with $L = W = B$, $2a/W \geq 0.6$, and notch thickness $t = 1$ mm is relatively reliable and optimal for the measurement of true mode-II fracture resistance The true mode-II fracture toughness for the single-edge notched SB specimen is determined from [23]:

$$
K_{IIc} = \frac{P(\sin\alpha - \tan\psi\cos\alpha)}{B\sqrt{W}} F_{II}
$$
$$
= \frac{P(\sin\alpha - \tan\psi\cos\alpha)}{B\sqrt{W}} \frac{2.138 - 5.2a/W + 6.674(a/W)^2 - 3.331(a/W)^3}{\sqrt{1 - a/W}}
$$

$$(3.6)$$

The true mode-II fracture toughness for the double-edge notched SB specimen is determined from [23–25]:

Fig. 3.6 Schematic of the SB specimen

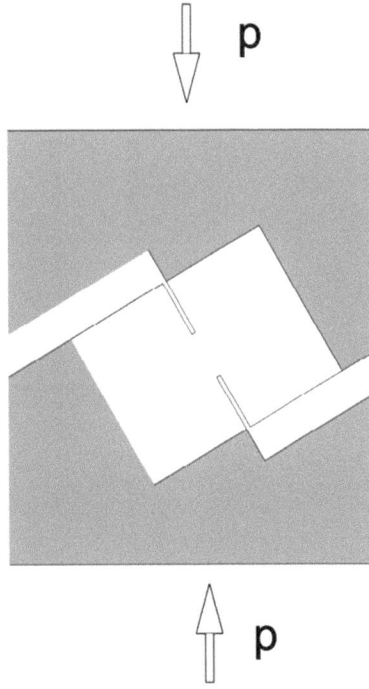

$$K_{IIc} = \frac{P(\sin\alpha - \tan\psi\cos\alpha)}{BW\sqrt{\pi a}^{-1}} F_{II}$$

$$= \frac{P(\sin\alpha - \tan\psi\cos\alpha)}{BW\sqrt{\pi a}^{-1}} \left[1.78 + 3.095\frac{2a}{W} - 10.559(\frac{2a}{W})^2 + 8.167(\frac{2a}{W})^3 \right]$$

$$(3.7)$$

where α is the precast notch inclination angle between the notch plane and the horizontal direction, which is suggested as $65 \sim 75°$, and ψ represents the internal friction angle of materials.

The third test specimen, known as the modified PTS (punch-through shear) specimen, is a cylinder of height H and diameter D, and contains the two symmetrical circular notches of diameter ID, depth d, and width w [29–31]. By applying the uniaxial compression loads, the circular rock bridge between the two notches can be broken, so the shear-induced fracture trajectory is self-planar or self-similar.

The schematic of the modified PTS specimen is displayed in Fig. 3.7. According to the recommendation of Yin et al. [31] and Yao et al. [30], the modified PTS specimen of height $H = 30$ mm and diameter $D = 50$ mm with the two symmetrical circular notches of diameter $ID = 25$ mm, depth $d = 10$ mm, and width $w = 1$ mm is relatively reliable and optimal for the measurement of true mode-II fracture resistance. The true mode-II fracture toughness for the modified PTS specimen specimen is determined from [31]:

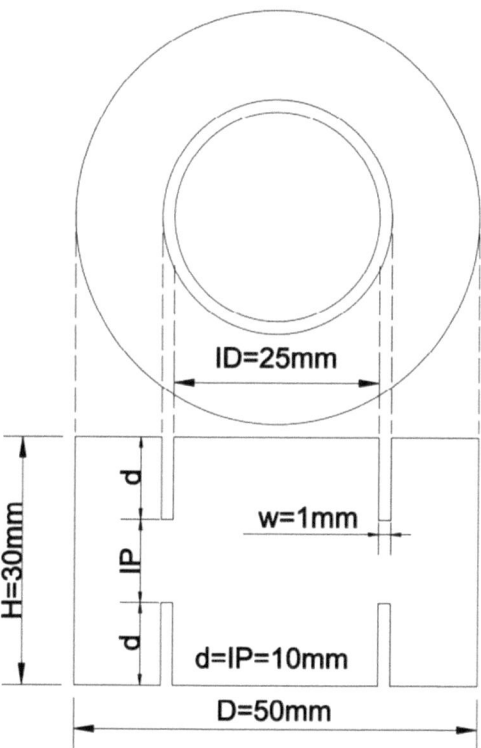

Fig. 3.7 Schematic of the modified PTS specimen

$$K_{IIc} = \eta \frac{P_{\max}}{A} = 0.03925 \frac{P_{\max}}{A} \tag{3.8}$$

where η represents the geometrical factor and is taken as 0.03925 m$^{0.5}$ [], and $A = 490.63$ mm^2 symbolizes the inner cross-sectional area of the modified PTS specimen.

The fourth test specimen, known as the ZCCDS (Z-shaped centrally cracked direct shear) specimen, is a cylinder of height H and diameter D, and contains a vertical straight-through notch of length $2a$ at the middle plane [32]. By applying the uniaxial compression loads, the direct shear stress is easily produced without any testing configurations. Accordingly, the self-planar cracking extension pattern can be achieved.

The schematic of the ZCCDS specimen is displayed in Fig. 3.8. According to the recommendation of Cao et al. [32], the ZCCDS specimen with $H/D = 1$, $l_1/H = 0.25$, and $2a/H = 0.3$ is relatively reliable and optimal for the measurement of true mode-II fracture resistance. The true mode-II fracture toughness for the SCC specimen is determined from [32]:

$$K_{IIc} = \frac{2P}{R^2}\sqrt{\frac{a}{\pi}}F_{II} = \frac{2P}{R^2}\sqrt{\frac{a}{\pi}}(3.577 - 36.974\frac{l_1}{H} - 14.97\frac{2a}{H} + 95.032(\frac{l_1}{H})^2 + 21.364(\frac{2a}{H})^2 + 108.4\frac{l_1}{H}\frac{2a}{H}) \tag{3.9}$$

Fig. 3.8 Schematic of the ZCCDS specimen

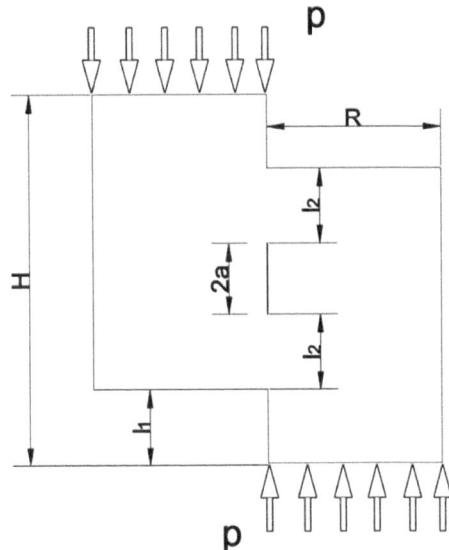

3.3 Test Methods of Mixed-Mode I/III Fracture

The first test specimen, known as the ENDB (edge-notched disk bending) specimen, is a disk of diameter D and thickness h and contains a straight edge notch of depth a and length D processed at the diametrical direction of the ENDB specimen. When conducting this testing procedure, the ENDB sample is supported by two symmetrical rollers with support span $2S$, and the upper roller is employed to produce the compressive force. Consequently, then the full range of mixed-mode I/III fracture can be accomplished depending on the loading angle, which is also regarded as the crack orientation angle between the prefabricated notch direction and the loading line. The mixed-mode I/III fracture parameters for the ENDB specimen are determined from [33–45]:

$$
\begin{cases}
K_I = \dfrac{6PS}{Rh^2}\sqrt{\pi a}F_I \\[2mm]
K_{III} = \dfrac{6PS}{Rh^2}\sqrt{\pi a}F_{III} \\[2mm]
K_e = \sqrt{K_I^2 + K_{III}^2} \\[2mm]
M^e = \dfrac{2}{\pi}\tan^{-1}(\dfrac{K_I}{K_{III}}) = \dfrac{2}{\pi}\tan^{-1}(\dfrac{F_I}{F_{III}}) \\[2mm]
T = \dfrac{6PS}{Rh^2}T^*
\end{cases}
\tag{3.10}
$$

The schematic of the ENDB specimen is displayed in Fig. 3.9a. Taking the ENDB specimen with $a/h = 0.6$ and $S/R = 0.925$ as an example, the variations of F_I, F_{II}, and T^* versus the loading angle β are outlined in Fig. 3.9b [46, 47]. Particularly, the loading angle $\beta = 0°$ corresponds to the pure mode-I loading case, while the loading angle $\beta = 62.5°$ corresponds to the pure mode-III loading case.

The second test specimen, known as the ENDC (edge-notched disk compression) specimen, is a disk of diameter D and thickness B and contains a straight edge notch of depth a and length D processed at the diametrical direction of the ENDC specimen, as illustrated in Fig. 3.10a. By applying diametrical compression loads, the complete range of mixed-mode I/III fracture can be accomplished depending on the loading angle, which is also regarded as the crack orientation angle between the prefabricated notch direction and the loading line. The mixed-mode I/III fracture parameters for the ENDC specimen are determined from [48–51]:

$$\begin{cases} K_I = \dfrac{P}{RB}\sqrt{\pi a}F_I \\[2mm] K_{III} = \dfrac{P}{RB}\sqrt{\pi a}F_{III} \\[2mm] K_e = \sqrt{K_I^2 + K_{III}^2} \\[2mm] M^e = \dfrac{2}{\pi}\tan^{-1}(\dfrac{K_I}{K_{III}}) = \dfrac{2}{\pi}\tan^{-1}(\dfrac{F_I}{F_{III}}) \end{cases} \tag{3.11}$$

The schematic of the ENDC specimen is displayed in Fig. 3.10a. Taking the ENDC specimen with $a/B = 0.6$ as an example, the variations of F_I and F_{II} versus the loading angle α are outlined in Fig. 3.10b [49]. Particularly, the loading angle $\beta = 0°$ corresponds to the pure mode-I loading case, while the loading angle $\beta = 11°$ corresponds to the pure mode-III loading case.

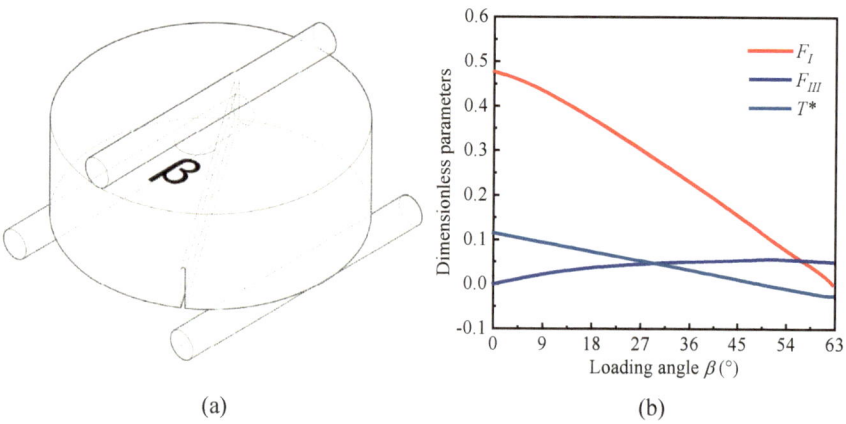

(a)   (b)

Fig. 3.9 Schematic of the ENDB specimen and dimensionless fracture parameters

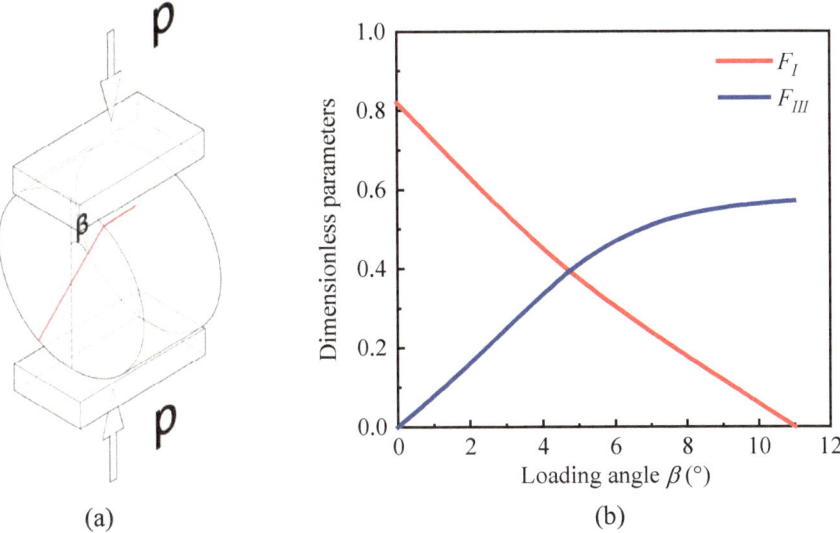

(a)   (b)

Fig. 3.10 Schematic of the ENDC specimen and dimensionless fracture parameters

The third test specimen, known as the DENDC (double-edge notched disk compression) specimen, is a disk of diameter D and thickness B and contains two straight edge notch of depth $a/2$ and length D processed at the diametrical direction of the DENDC specimen, as illustrated in Fig. 3.11a. By applying diametrical compression loads, the complete range of mixed-mode I/III fracture can be accomplished depending on the loading angle, which is also regarded as the crack orientation angle between the prefabricated notch direction and the loading line. The mixed-mode I/III fracture parameters for the DENDC specimen are determined from [49]:

$$
\begin{cases}
K_I = \dfrac{P}{RB}\sqrt{\pi\dfrac{a}{2}}F_I \\[2mm]
K_{III} = \dfrac{P}{RB}\sqrt{\pi\dfrac{a}{2}}F_{III} \\[2mm]
K_e = \sqrt{K_I^2 + K_{III}^2} \\[2mm]
M^e = \dfrac{2}{\pi}\tan^{-1}\left(\dfrac{K_I}{K_{III}}\right) = \dfrac{2}{\pi}\tan^{-1}\left(\dfrac{F_I}{F_{III}}\right)
\end{cases}
\tag{3.12}
$$

The schematic of the DENDC specimen is displayed in Fig. 3.11a. Taking the DENDC specimen with $a/B = 0.6$ as an example, the variations of F_I and F_{II} versus the loading angle α are outlined in Fig. 3.11b [49]. Particularly, the loading angle $\beta = 0°$ corresponds to the pure mode-I loading case, while the loading angle $\beta = 27°$ corresponds to the pure mode-III loading case.

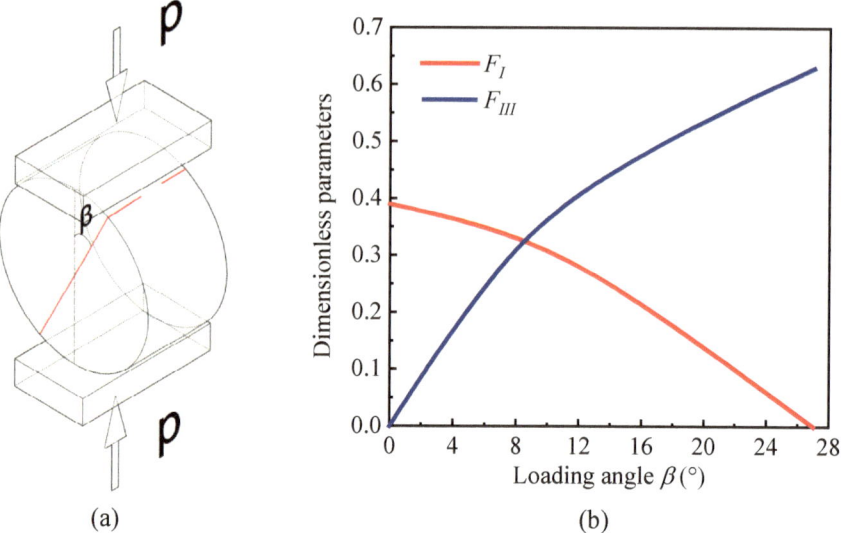

Fig. 3.11 Schematic of the DENDC specimen and dimensionless fracture parameters

The forth test specimen, known as the TCSCB (tilt-cracked semi-circle bending) specimen, is a disk of diameter D and thickness B and contains a tilted edge crack of depth a and tilted angle α processed at the bottom of the TCSCB specimen, as illustrated in Fig. 3.12a. Using the straightforward three-point bend configuration, a limited number of mixed-mode I/III fracture can be accomplished depending on the angular position of the tilted edge crack. The mixed-mode I/III fracture parameters for the TCSCB specimen are determined from [52]:

$$
\begin{cases}
K_I = \dfrac{P}{2RB}\sqrt{\pi a}F_I \\[2mm]
K_{III} = \dfrac{P}{2RB}\sqrt{\pi a}F_{III} \\[2mm]
K_e = \sqrt{K_I^2 + K_{III}^2} \\[2mm]
M^e = \dfrac{2}{\pi}\tan^{-1}(\dfrac{K_I}{K_{III}}) = \dfrac{2}{\pi}\tan^{-1}(\dfrac{F_I}{F_{III}})
\end{cases}
\tag{3.13}
$$

The schematic of the TCSCB specimen is displayed in Fig. 3.12(a). Taking the DENDC specimen with radius of $R = 75$ mm, thickness of $B = 32$ mm, crack depth of $a = 20$ mm, and support span of $2S = 120$ mm as an example, the variations of F_I and F_{III} versus the tilted angle α are outlined in Fig. 3.12b [52]. Particularly, the tilted angle $\beta = 0°$ corresponds to the pure mode-I loading case, while the tilted angle $\beta = 60°$ corresponds to the mixed-mode I/III loading case. Note that this TCSCB specimen fails to provide the pure mode-III and predominant mode-III loadings,

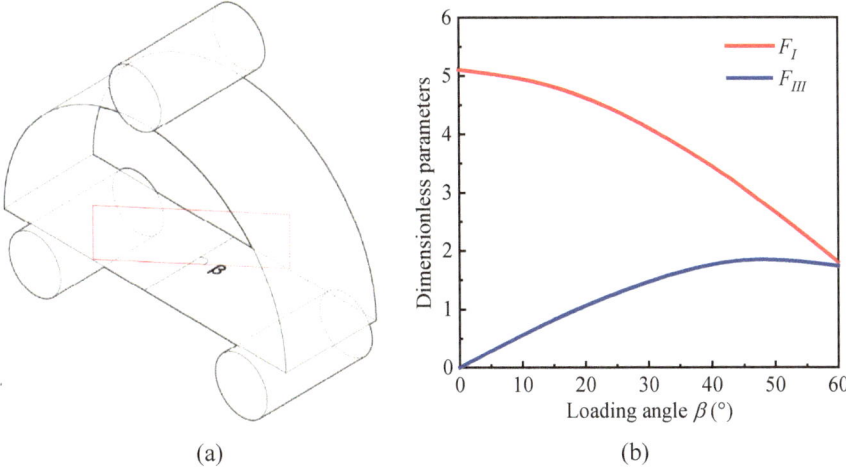

(a)   (b)

Fig. 3.12 Schematic of the TCSCB specimen and dimensionless fracture parameters

indicating that this TCSCB specimen is suitable for assessment of mixed-mode I/III fracture resistance for the case of $M^e > 0.5$.

The fifth test specimen, known as the ATCSCB (antisymmetrical tilt-cracked semi-circle bending) specimen, is a disk of diameter D and thickness B and contains a inclined edge crack of depth a and tilted angle α processed at the bottom of the ATCSCB specimen, as illustrated in Fig. 3.13a. Using the antisymmetrical three-point bend configuration, a limited number of mixed-mode I/III fractures can be accomplished by changing the angular position of the inclined edge crack. The mixed-mode I/III fracture parameters for the ATCSCB specimen are determined from [53]:

$$
\begin{cases}
K_I = \dfrac{P}{2RB}\sqrt{\pi a}F_I \\[2mm]
K_{III} = \dfrac{P}{2RB}\sqrt{\pi a}F_{III} \\[2mm]
K_e = \sqrt{K_I^2 + K_{III}^2} \\[2mm]
M^e = \dfrac{2}{\pi}\tan^{-1}(\dfrac{K_I}{K_{III}}) = \dfrac{2}{\pi}\tan^{-1}(\dfrac{F_I}{F_{III}})
\end{cases}
\tag{3.14}
$$

The schematic of the ATCSCB specimen is displayed in Fig. 3.13a. Taking the ATCSCB specimen with diameter of $R = 95$ mm, thickness of $B = 40$ mm, crack depth of $a = 22$ mm, and support span of $2S = 76$ mm as an example, the variations of F_I and F_{III} versus the inclined angle α are outlined in Fig. 3.13b [53]. Particularly, the tilted angle $\beta = 90°$ corresponds to the pure mode-III loading state, the inclined angle $\beta = 60°$ corresponds to the mixed-mode I/III loading state. Note that this TCSCB specimen fails to provide the pure mode-I loading.

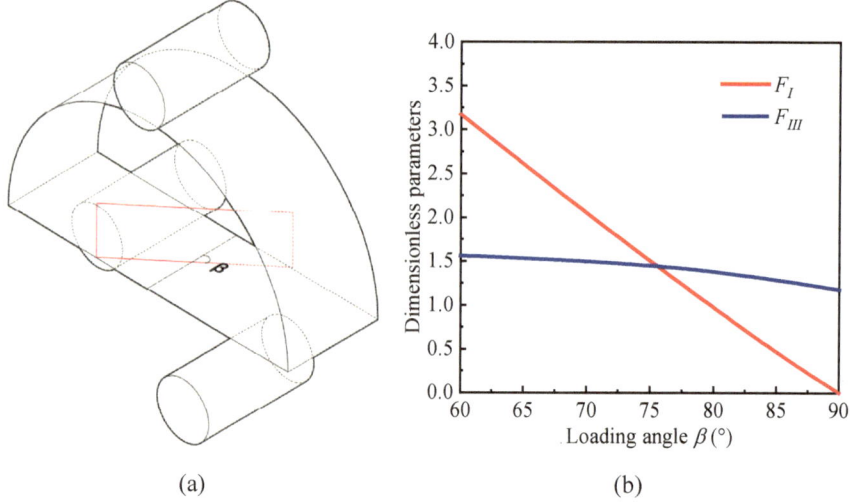

(a)   (b)

Fig. 3.13 Schematic of the ATCSCB specimen and dimensionless fracture parameters

References

1. Ayatollahi MR, Sistaninia M (2011) Mode II fracture study of rocks using Brazilian disk specimens. Int J Rock Mech Min 48:819–826
2. Ayatollahi MR, Aliha MRM (2009) Mixed mode fracture in soda lime glass analyzed by using the generalized MTS criterion. Int J Solids Struct 46:311–321
3. Ayatollahi MR, Aliha MRM (2007) Wide range data for crack tip parameters in two disc-type specimens under mixed mode loading. Comput Mater Sci 38:660–670
4. Aliha MRM, Ayatollahi MR (2009) Brittle fracture evaluation of a fine grain cement mortar in combined tensile-shear deformation. Fatig Fract Eng Mater Struct 32:987–994
5. Cao RH, Wang CS, Yao RB, Hu T, Lei DX, Lin H, Zhao YL (2020) Effects of cyclic freeze-thaw treatments on the fracture characteristics of sandstone under different fracture modes: laboratory testing. Theor Appl Fract Mec 109:102738
6. Kang P, Hong Lv, Fazhi Y, Quanle Z, Xiao S, Zhaopeng L (2020) Effects of temperature on mechanical properties of granite under different fracture modes. Eng Fract Mech 226:106838
7. Ren L, Xie LZ, Xie HP, Ai T, He B (2016) Mixed-mode fracture behavior and related surface topography feature of a typical sandstone, Rock. Mech. Roc.k Eng. 49 (2016) 3137–3153.
8. Ren Li, Zhu Z, Wang M, Zheng T, Ai T (2014) Mixed-mode elastic-plastic fractures: improved r-criterion. J Eng Mech 140(6):04014033
9. Luo Y, Ren L, Xie LZ, Ai T, He B (2017) Fracture behavior investigation of a typical sandstone under mixed-mode I/II loading using the notched deep beam bending method. Rock Mech Rock Eng 50:1987–2005
10. Omidvar N, Aliha MRM, Khoramishad H (2023) Hygrothermal degradation of MWCNT/epoxy brittle materials under I/II combined mode loading conditions: An experimental, micro structural and theoretical study. Theoret Appl Fract Mech 125:103896
11. Aliha MRM, Samareh-Mousavi SS, Mirsayar MM (2021) Loading rate effect on mixed mode I/II brittle fracture behavior of PMMA using inclined cracked SBB specimen. Int J Solids Struct 232:111177
12. Mousavi SS, Aliha MRM, Imani DM (2020) On the use of edge cracked short bend beam specimen for PMMA fracture toughness testing under mixed-mode I/II. Polym Testing 81:106199

13. Hua W, Zhu Z, Zhang W, Li J, Huang J, Dong S Application of modified fracture criteria incorporating T-stress for various cracked specimens under mixed mode I-II loading
14. Shen Z, Yu HJ, Guo LC, Hao LL, Huang K (2022) A modified G criterion considering T-stress and differentiating the separation and shear failure in crack propagation. Int J Solids Struct 236–237:111357
15. Aliha MRM, Hosseinpour GhR, Ayatollahi MR (2013) Application of cracked triangular specimen subjected to three-point bending for investigating fracture behavior of rock materials. Rock Mech Rock Eng 46:1023–1034
16. Ying Xu, Yao W, Zhao G, Xia K (2020) Evaluation of the short core in compression (SCC) method for measuring mode II fracture toughness of rocks. Eng Fract Mech 224:106747
17. Li D, Zhang C, Zhu Q, Ma J, Gao F (2022) Deformation and fracture behavior of granite by the short core in compression method with 3D digital image correlation. Fatigue Fract Eng Mater Struct 45:425–440
18. Li X, Ying Xu, Zhan Z, Yao W, Xia K (2022) Influence of thermal treatment on dynamic mode II fracture properties of rocks using the short core in compression (SCC) method. Theoret Appl Fract Mech 119:103383
19. Zhang C, Li D, Wang C, Ma J, Zhou A, Xiao P (2022) Effect of confining pressure on shear fracture behavior and surface morphology of granite by the short core in compression test. Theoret Appl Fract Mech 121:103506
20. Alneasan M (2023) Abdel Kareem Alzo'ubi, Nader Okasha, A comprehensive study for the effect of sample geometry and lateral pressure on shear fractures using the short core in compression (SCC) method. European Journal of Mechanics/A Solids 100:104988
21. Zhang C, Li D, Ma J, Zhu Q, Luo P, Chen Y, Han M (2023) Dynamic shear fracture behavior of rocks: insights from three-dimensional digital image correlation technique. Eng Fract Mech 277:109010
22. Ri-hong C, Fang L, Qiu X, Lin H, Li X, Li W, Qiao Q (2023) Effect of heating–water cooling cycle treatment on the pore structure and shear fracture characteristics of granite 109263
23. Rao Q, Sun Z, Stephansson O, Li C, Stillborg B (2003) Shear fracture (Mode II) of brittle rock. Int J Rock Mech Mining Sci 40:355–375
24. Sun D, Rao Q, Wang S, Shen Q, Yi W Shear fracture (Mode II) toughness measurement of anisotropic rock
25. Cao R-H, Wang C, Tao Hu, Yao R, Li T, Lin Q (2022) Experimental investigation of plane shear fracture characteristics of sandstone after cyclic freeze–thaw treatments. Theoret Appl Fract Mech 118:103214
26. Fan ZD, Xie HP, Zhang R, Lu HJ, Zhou Q, Nie XF, Luo Y, Ren L (2022) Characterization of anisotropic mode II fracture behaviors of a typical layered rock combining AE and DIC techniques. Eng Fract Mech 271:108599
27. Fan ZD, Ren L, Xie HP, Zhang R, Li CB, Lu HJ, Zhang AL, Zhou Q, Ling WQ (2023) 3D anisotropy in shear failure of a typical shale. Petroleum Sci 20:212−229
28. Fan ZD, Xie HP, Ren L, Zhang R, He R, Li CB, Zhang ZT, Wang J, Xie J (2022) Anisotropy in shear-sliding fracture behavior of layered shale under different normal stress conditions. J Cent South Univ 29(11):3678–3694
29. Yao W, Wang JX, Wu BB, Xu Y, Xia KW (2023) Dynamic mode II fracture toughness of rocks subjected to various in situ stress conditions. Rock Mech Rock Eng 56:2293–2310
30. Yao W, Xu Y, Xia KW, Wang S (2020) Dynamic mode II fracture toughness of rocks subjected to confining pressure. Rock Mech Rock Eng 53:569–586
31. Yin TB, Tan XS, Wu Y, Yang Z, Li MJ (2021) Temperature dependences and rate effects on Mode II fracture toughness determined by punch-through shear technique for granite. Theor Appl Fract Mech 114:103029
32. Cao P, Zhou T, Zhu J (2023) A novel testing method for examining mode II fracture of rock and its application 109831
33. Shi ZM, Li JT, Wang MX, Tan H, Lin H, Li KH (2023) Effect of temperature on pure mode III fracture behavior and fracture morphology of granite after thermal shock. Theor Appl Fract Mech 127:104024

34. Zhao Y, Tang W, Zhang Y, et al. (2024) Evaluating fracture resistance of basalt fiber reinforced mortar to mode I/III load using edge notched disc bend (ENDB) specimen: Insights from acoustic emission and morphological analysis. Constr Build Mater (411):134421

35. Bidadi J, Akbardoost J, Aliha MRM (2019) Thickness effect on the mode III fracture resistance and fracture path of rock using ENDB specimens. Fatigue Fract Eng Mater Struct 43:277–291

36. Yang Z, Yin T, Zhuang D, et al. (2022) Effect of temperature on mixed mode I/III fracture behavior of diorite: An experimental investigation. Theor Appl Fract Mech (122):103571

37. Shen Z, Yu HY, Guo LC, Hao LL, Zhu S, Huang K (2023) A modified 3D G-criterion for the prediction of crack propagation under mixed mode I-III loadings. Eng Fract Mech 281:109082

38. Liu Z, Ma C, Wei X (2023) Study on fracture behavior of layered limestone under mixed mode I/III loading 128:104102

39. Liu Z, Ma C, Wei X (2024) Assessment of mode I/III fracture toughness of bi-material rock-like ENDB and ENDC specimens 129:104235

40. Cao R-H, Yao R, Dai H, Qiu X, Lin H, Li K (2024) Fracture behaviour of transversely isotropic rocks under pure mode III fracture: Experiment and numerical simulation 129:104208

41. Aliha MRM, Mousavi SS, Bahmani A et al (2019) Crack initiation angles and propagation paths in polyurethane foams under mixed modes I/II and I/III loading. Theor Appl Fract Mech 101:152–161

42. Shahbazian B, Mirsayar MM, Aliha MRM, Darvish MG, Asadi MM, Haghighatpour PJ (2022) Experimental and theoretical investigation of mixed-mode I/II and I/III fracture behavior of PUR foams using a novel strain-based criterion. Int J Solids Struct 258:111996

43. Pietras D, Aliha MRM, Kucheki HG, Sadowski T (2023) Tensile and tear-type fracture toughness of gypsum material: direct and indirect testing methods. J Rock Mech Geotech Eng 15:1777–1796

44. Aliha MRM, Kosarneshan K, Salehi SM, Haghighatpour PJ, Mousavi A (2023) On the Statistical prediction of K_{Ic} and G_{Ic} for railway andesite ballast rock using different three-point bend disc samples. Rock Mech Rock Eng 56:5181–5202

45. Pirmohammad S, Bayat A (2016) Characterizing mixed mode I/III fracture toughness of asphalt concrete using asymmetric disc bend (ADB) specimen. Const Build Mater 120:571–580

46. Aliha MRM, Bahmani A (2017) Rock fracture toughness study under mixed mode I/III loading. Rock Mech Rock Eng 50:1739–1751

47. Aliha MRM, Bahmani A, Akhondi S (2015) Numerical analysis of a new mixed mode I/III fracture test specimen. Eng Fract Mech 134:95–110

48. Bahmani A, Farahmand F, Janbaz MR, Darbandi AH, Ghesmati-Kucheki H, Aliha MRM (2021) On the comparison of two mixed-mode I + III fracture test specimens. Eng Fract Mech 241:107434

49. Aliha MRM, Kucheki HG, Asadi MM (2021) On the use of different diametral compression cracked disc shape specimens for introducing mode III deformation. Fatig Fract Eng Mater Struct 1–17

50. Karimi HR, Bidadi J, Aliha MRM, Mousavi A (2023) Mohammadi MH, Haghighatpour PJ (2023) An experimental study and theoretical evaluation on the effect of specimen geometry and loading configuration on recorded fracture toughness of brittle construction materials. J Build Eng 75:106759

51. Aliha MRM, Sarbijan MJ, Bahmani A (2017) Fracture toughness determination of modified HMA mixtures with two novel disc shape configurations. Constr Build Mater 155:789–799

52. Pirmohammad S, Kiani A (2016) Study on fracture behavior of HMA mixtures under mixed mode I/III loading. Eng Fract Mech 153:80–90

53. Bakhshizadeh M, Pirmohammad S (2022) Experimental and numerical evaluation of semi-circular bending specimen for mixed mode I/III and pure mode III fracture tests. Fatigue Fract Eng Mater Struct 45:1213–1226

Chapter 4
Mixed-Mode I/II Fracture

As the inherent nature of rocks, natural cracks play a remarkable part in controlling the mechanics and permeability responses in rock masses. Due to the intense stress concentration at their neighborhoods, these cracks are extensively recognized as the initial locations for the initiation, extension, and convergence of cracking [1–10]. The fracture toughness (i.e., critical stress intensity factor) plays an important role in rock fracture mechanics, which weighs the stress and displacement fields near the notch tip, ultimately governing the stability of rock engineering and the exploitation of geoenergy [11–20]. In fact, the engineering rock masses generally subjected to complicated combined-mode I/II loadings, the resultant fracturing problems need to be investigated thoroughly [21–30].

A SCB (semi-circular bending) specimen under symmetrical three-point bending has been popularly deployed in brittle and quasi-brittle materials because of its capability to permit brittle fractures from pure tension (i.e., mode I) to pure shear (i.e., mode II) [31]. Using the SCB test method, Ayatollahi and Akbardoost [32] investigated the specimen size effect on modes I and II fracture toughness of marble. The measured results showed that modes I and II fracture resistances increase with the increasing specimen sizes, and the variations in fracture toughness measurements can be theoretically explained using the modified MTS (maximum tangential stress) criterion [33]. Aliha et al. [12] adopted the CCBD and SCB test methods to evaluate and compare the combined-mode I/II fracture resistance of marble, the experimental results indicated that the upper boundary combined-mode I/II fracture resistance envelope can be obtained by the CCBD testing, the lower boundary combined-mode I/II fracture resistance envelope can be determined from the SCB testing. Based on the generalized MTS criterion, negative T-stresses in the CCBD specimen can increase the combined-mode I/II fracture toughness, and conversely positive T-stresses in the SCB specimen can decrease it. For the SCB marble specimens with different notch tip shapes, the interval and tip shape of the prefabricated notch should be processed as narrow and sharp to accomplish a more precise measurement of rock mode-I fracture toughness [34]. The mode-II fracture resistance was estimated and compared by

© The Author(s) 2024
Y. Zhao et al., *Rock Fracture Mechanics and Fracture Criteria*,
https://doi.org/10.1007/978-981-97-5822-7_4

Bahrami et al. [35] and Pirmohammad et al. [36] using the SCB testing fixture with three distinct types of supports (i.e., roller, roller-in-groove, and fixed supports). Since there are negligible friction forces at the contact zones between the roller supports and the SCB specimen, the mode-II fracture toughness measurements obtained using the roller-type supports are relatively dependable. Compared with another SCB test specimen with a chevron-type notch, the SCB test specimen with a straight-through notch is an expedient measure for the evaluation of mode-I fracture toughness [37]. One plausible explanation is that (1) the abovementioned test methods can provide equivalent mode-I fracture resistance values and (2) the SCB test specimen with a straight-through notch can be readily machined from rock masses.

4.1 Experiment Apparatuses and Specimen Preparation

The combined-mode I/II fracturing experiments are conducted on the SCB specimens by the DANA mechanical tester with a testing capacity of 100 kN, and the test procedure for the mixed-mode I/II fracture is displayed in Fig. 4.1. As previously documented, the lower loading rates had negligible influences on the fracture resistance measurements. To properly eliminate the loading rate effect on the fracture resistance measurements, a lower loading rate should be employed. Particularly, a relatively constant fracture resistance is expected to be determined since it is recognized as an independently inherent material property [38, 39]. In the current study, the SCB specimens of sandstone are loaded monotonously at a lower testing rate of 0.1 mm/min.

Fig. 4.1 Test procedure for the mixed-mode I/II fracture

Table 4.1 Variations of nondimensional mixed-mode I/II fracture indicators (i.e., F_I, F_{II}, and T^*) versus the loading angle for $a/R = 0.4$ and $S/R = 0.4$

Loading mode	Loading angle β (°)	T^*	Mode-I geometry factor F_I	Mode-II geometry factor F_{II}
Pure mode I	0	−1.26	2.17	0
Mixed-mode I/II	15	−0.40	1.56	0.78
Mixed-mode I/II	25	0.53	0.89	1.05
Pure mode II	35	1.62	0	1.15

These sandstone blocks are first machined as the semi-disk specimens without preset notches, their dimensions are designed as $2R = D = 75$ mm in diameter and $B = 25$ mm in thickness based on the investigation of Ren et al. [40]. Then the straight-through notch with the length of $a = 15$ mm and the interval of 0.5 mm is prefabricated in the semi-disk specimen bottom along the diametrical orientation. The combined-mode I/II fracturing tests are implemented by using the representative specimens with specific notch inclination angles between the notch plane and the vertical direction (i.e., loading direction), and the loading angles are selected as 0° (i.e., pure mode-I loading), 15° (i.e., mixed-mode I/II loading), 25° (i.e., mixed-mode I/II loading), and 35° (i.e., pure mode-II loading). According to the investigated results of Ren and co-workers [], the variations of nondimensional mixed-mode I/II fracture indicators (i.e., F_I, F_{III}, and T^*) versus the loading angle (i.e., α) for $a/R = 0.4$ and $S/R = 0.4$ is described in Table. 4.1 [40].

4.2 Experimental Results and Analyses

4.2.1 Evaluation of Fracture Load and Fracture Toughness

Figure 4.2 shows the developmental curves of diametrical compression load versus loading-point deflection for SCB specimens of sandstone under different combined-mode I/II loadings. Obviously, the curve shapes for each tested SCB specimen are characterized by three deformation stages: (1) initial nonlinearity, (2) linear elasticity with the elevated force, and (3) ultimate brittleness. One can conclude from Fig. 4.3 that the magnitude of the fracture load P_{max} (i.e., peak load) for the SCB sandstone specimens is improved by transitioning from pure mode-I loading to pure mode-II loading. When the mode mixity index $M^e = 1, 0.71, 0.45$, and 0, the average P_{max} values are respectively 2336, 2553, 3097, and 3697 N. The critical SIFs can be computed by introducing the values of P_{max}, F_I, and F_{II} into Eq. (3.2), as plotted in Fig. 4.4. When the mode mixity index $M^e = 1, 0.71, 0.45$, and 0, the average K_e values are respectively 0.59, 0.51, 0.49, and 0.49 MPa·m$^{1/2}$. Consequently, the tested ratio $K_{IIc}/K_{Ic} = 0.84$. It can be observed from Fig. 4.4 that the magnitude of the effective fracture toughness K_e depends on the mode mixity indicator M^e and is decreased

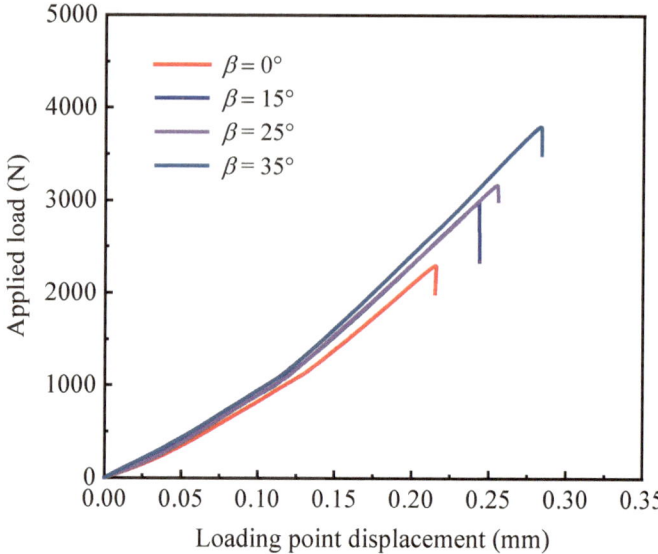

Fig. 4.2 Typical load–displacement curves for SCB sandstone specimens

by transitioning from pure mode-I loading to pure mode-II loading. However, this opposite phenomenon is observed in the CCBD testing [41]. This indicates that the combined-mode I/II fracture resistance is dependent on the geometry and loading configurations of test specimens.

4.2.2 Determination of Fracture Progress Zone Radius and Fracture Initiation Angle

As documented previously, the size for the damage zone (i.e., fracture progress zone) is recognized as a constant material parameter for combined-mode fracture problems. According to the suggestion of Aliha and co-workers [38, 39], the magnitude for this zone can be assessed as

$$r_c = \frac{1}{2\pi}(\frac{K_{Ic}}{\sigma_t})^2 \tag{4.1}$$

where σ_t represents the tension strength. Using the Brazilian disc testing method, the mean σ_t value is determined as 3.50 MPa.

The fracture initiation angle θ_c is defined as the angle between the preset notch plane and the initial fracturing direction, as provided in Fig. 4.5 [42]. According to the abovementioned approach to judging this angle, the angles for the fractured sandstone

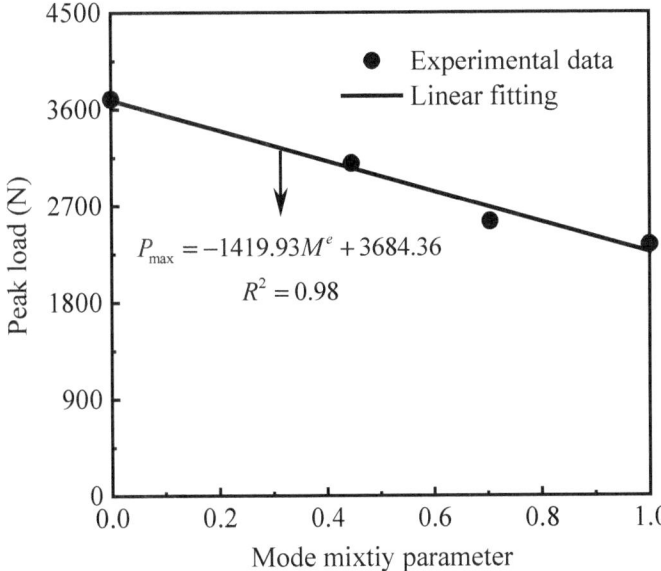

Fig. 4.3 Variations of the fracture load P_{max} versus the mode mixtiy index for tested SCB sandstone specimens

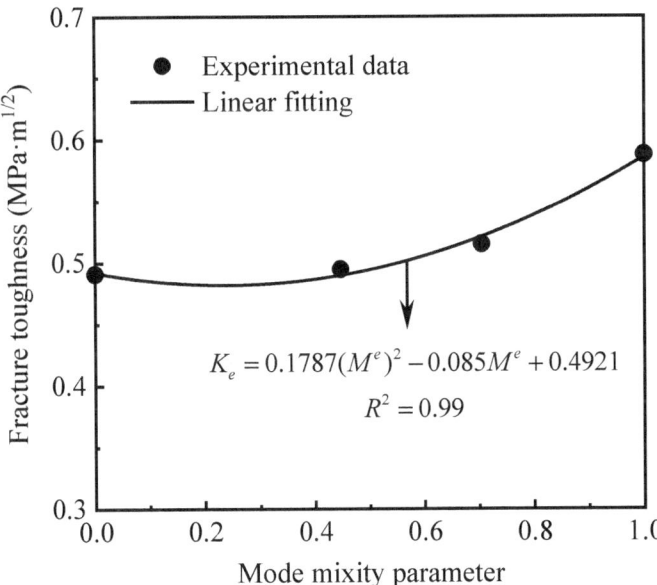

Fig. 4.4 Variations of the effective fracture toughness K_e versus the mode mixity index for tested SCB sandstone specimens

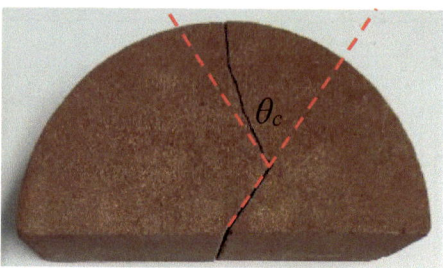

Fig. 4.5 Testing method of the fracture initiation angle θ_c

(1) $\alpha = 0°$ (pure mode I)   (2) $\alpha = 15°$ (3) $\alpha = 25°$ (4) $\alpha = 35°$ (pure mode II)

Fig. 4.6 Fractured SCB sandstone specimens

specimens (see Fig. 4.6) are manually measured. By transitioning from pure mode-I loading to pure mode-II loading, the macroscopic fracture trajectories vary from self-planar extension patterns to curvilinear-type ones. According to the conclusion of Zheng and co-workers [43], the negative T-stresses can stabilize the fracture path in a self-planar propagation manner due to their passive contribution to the tangential stress, resulting in the reduced fracture kinking angle. It can be observed from Fig. 4.7 that the prediction accuracy of the GMTS (generalized maximum tangential stress) fracture criterion is dependent on the fracture progress zone size r_c. Note that r_c is calculated as 4.53 mm from Eq. (4.1), and r_c is assumed as 0.01 mm in this work. Hence, the r_c value should be properly selected when the GMTS criterion is employed to evaluate and predict the fracture initiation angle θ_c.

4.2.3   Application of Fracture Criteria

Figure 4.8 illustrates the comparisons of combined-mode I/II fracture toughness ratio between the experimental results and the theoretical predictions. When the fracture progress zone size r_c is assumed as 0.01 mm, the combined-mode I/II fracture toughness envelope predicted by the GMTS criterion is similar to that obtained by the MTS criterion. When the fracture progress zone size r_c is taken as 4.53 mm, the GMTS criterion fails to provide accurate predictions for the combined-mode I/II fracture resistance of tested SCB sandstone specimens. Hence, the reasonable selection of the fracture progress zone size r_c plays an important role in improving the prediction accuracy of the GMTS fracture model. Aliha et al. [12] suggested

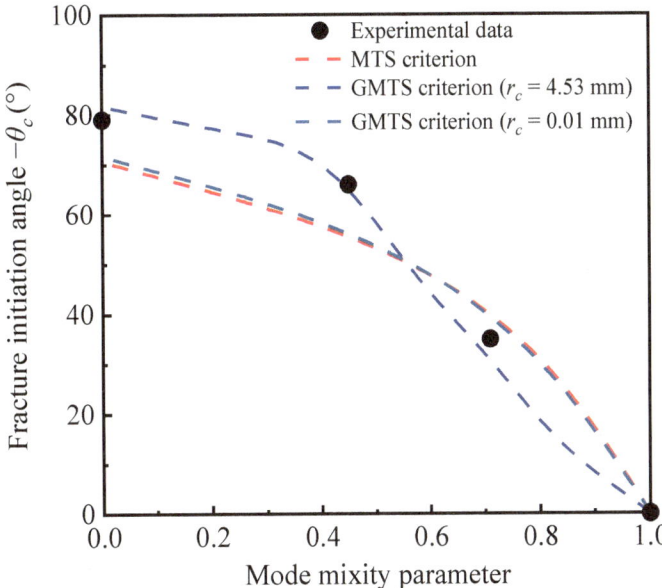

Fig. 4.7 Comparisons in fracture initiation angle between the measured and theoretical results

that the combined-mode I/II fracture resistance envelope determined from the SCB testing method could be recognized as the benchmark of the lower boundary for engineering applications.

4.3 Verification of a Modified Average Distortional Strain Energy Density Criterion

Considering the singular terms of Williams' series expansion [44], the plane stress field near the notch tip in a Cartesian coordinate system can be described as

$$
\begin{cases}
\sigma_{xx} = \dfrac{K_I}{\sqrt{2\pi r}}\cos\dfrac{\theta}{2}\left(1 - \sin\dfrac{\theta}{2}\sin\dfrac{3\theta}{2}\right) - \dfrac{K_{II}}{\sqrt{2\pi r}}\sin\dfrac{\theta}{2}\left(2 + \cos\dfrac{\theta}{2}\cos\dfrac{3\theta}{2}\right) \\[3mm]
\sigma_{yy} = \dfrac{K_I}{\sqrt{2\pi r}}\cos\dfrac{\theta}{2}\left(1 + \sin\dfrac{\theta}{2}\sin\dfrac{3\theta}{2}\right) + \dfrac{K_{II}}{\sqrt{2\pi r}}\sin\dfrac{\theta}{2}\cos\dfrac{\theta}{2}\cos\dfrac{3\theta}{2} \\[3mm]
\sigma_{xy} = \dfrac{K_I}{\sqrt{2\pi r}}\sin\dfrac{\theta}{2}\cos\dfrac{\theta}{2}\cos\dfrac{3\theta}{2} + \dfrac{K_{II}}{\sqrt{2\pi r}}\cos\dfrac{\theta}{2}\left(1 - \sin\dfrac{\theta}{2}\sin\cos\dfrac{3\theta}{2}\right) \\[3mm]
\sigma_{zz} = v'(\sigma_{xx} + \sigma_{yy})
\end{cases}
\tag{4.2}
$$

According to Eq. (4.2), the two-dimensional distortional strain energy density W_d can be defined as follows [45]:

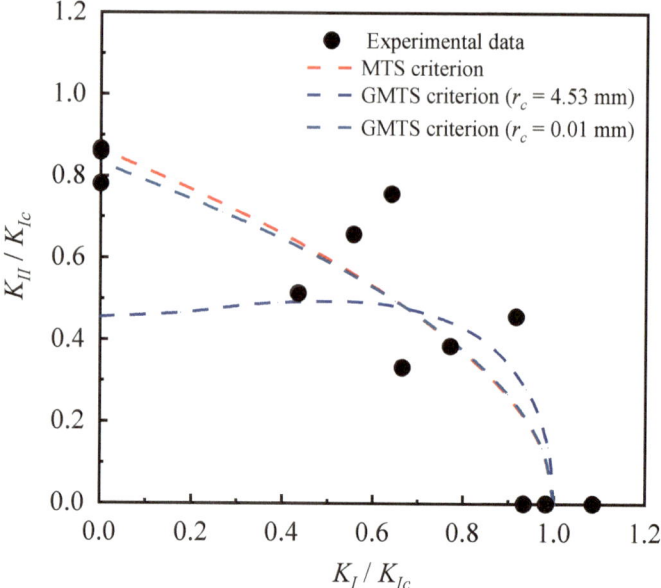

Fig. 4.8 Comparison in combined-mode I/II fracture resistance between the experimental and theoretical results

$$W_d = \frac{1+\nu}{6E}\left[(\sigma_{xx}-\sigma_{yy})^2 + (\sigma_{yy}-\sigma_{zz})^2 + (\sigma_{xx}-\sigma_{zz})^2 + 6\sigma_{xy}^2\right] \tag{4.3}$$

Using the integral method, the distortional strain energy in the circular plastic zone with a critical radius r_c from the notch tip can be expressed as

$$E_d(r_c) = \int_0^{r_c}\int_{-\pi}^{\pi} W_d r dr d\theta \tag{4.4}$$

Thus, the average distortional strain energy density $\overline{E_d(r_c)}$ in the circular plastic zone can be written as

$$\overline{E_d(r_c)} = \frac{E_d(r_c)}{\pi r_c^2} = \frac{7-16\nu'+16\nu'^2}{48\pi G r_c}K_I^2 + \frac{19-16\nu'+16\nu'^2}{48\pi G r_c}K_{II}^2 \tag{4.5}$$

Under the pure mode-I fracture, there are $r_c = r_{Ic}$, $K_I = K_{Ic}$, and $K_{II} = 0$, then the Eq. (4.5) can be simplified as

$$\overline{E_{dc}(r_c)} = \frac{7-16\nu'+16\nu'^2}{48\pi G r_{Ic}}K_{Ic}^2 \tag{4.6}$$

Under the pure mode-II fracture, there are $r_c = r_{IIc}$, $K_I = 0$, and $K_{II} = K_{IIc}$, substituting Eq. (4.6) into Eq. (4.5) yields

$$\frac{K_{IIc}}{K_{Ic}} = \sqrt{\frac{7 - 16\nu\prime + 16\nu\prime 2}{19 - 16\nu\prime + 16\nu\prime 2} \frac{r_{IIc}}{r_{Ic}}} \tag{4.7}$$

Under the general combined-mode I/II fracture, we consider the effective critical plastic zone radius r_{ec} instead of the conventional invariable radius r_{Ic} to improve the traditional average distortional strain energy density criterion. Substituting Eq. (4.6) into Eq. (4.5) yields

$$\frac{7 - 16\nu\prime + 16\nu\prime 2}{r_{Ic}} K_{Ic}^2 = \frac{7 - 16\nu\prime + 16\nu\prime 2}{r_{ec}} K_I^2 + \frac{19 - 16\nu\prime + 16\nu\prime 2}{r_{ec}} K_{II}^2 \tag{4.8}$$

In the current study, the classical linear interpolation method is adopted to determine the effective critical plastic zone radius r_{ec}:

$$r_e = \frac{M^e - 0}{1 - 0} r_{Ic} - \frac{M^e - 1}{1 - 0} r_{IIc} = M^e r_{Ic} - (M^e - 1) r_{IIc} \tag{4.9}$$

Substituting Eq. (4.9) into Eq. (4.8) yields

$$(7 - 16\nu\prime + 16\nu\prime 2) K_{Ic}^2 = \frac{7 - 16\nu\prime + 16\nu\prime 2}{M^e - (M^e - 1)\frac{r_{IIc}}{r_{Ic}}} K_I^2 + \frac{19 - 16\nu\prime + 16\nu\prime 2}{M^e - (M^e - 1)\frac{r_{IIc}}{r_{Ic}}} K_{II}^2 \tag{4.10}$$

When both sides of Eq. (4.10) are divided by K_I^2, K_{II}^2, , respectively, the normalized fracture toughness (i.e., $K_I = K_{Ic}$ and $K_{II} = K_{Ic}$) can be extracted to evaluate and predict the onset of combined-mode I/II fracture as follows:

$$\begin{cases} \dfrac{K_I}{K_{Ic}} = \left\{ \dfrac{1}{M^e - (M^e - 1)\frac{r_{IIc}}{r_{Ic}}} + \dfrac{19 - 16\nu\prime + 16\nu\prime 2}{(7 - 16\nu\prime + 16\nu\prime 2)[M^e - (M^e - 1)\frac{r_{IIc}}{r_{Ic}}]} \dfrac{K_{II}^2}{K_I^2} \right\}^{-0.5} \\[2em] \dfrac{K_{II}}{K_{Ic}} = \left\{ \dfrac{1}{M^e - (M^e - 1)\frac{r_{IIc}}{r_{Ic}}} \dfrac{K_I^2}{K_{II}^2} + \dfrac{19 - 16\nu\prime + 16\nu\prime 2}{(7 - 16\nu\prime + 16\nu\prime 2)[M^e - (M^e - 1)\frac{r_{IIc}}{r_{Ic}}]} \right\}^{-0.5} \end{cases} \tag{4.11}$$

where M^e indicates the mode mixity index. For the plane stress case, $\nu\prime = 0$; for the plane strain case, $\nu\prime = \nu$.

According to the 2D-MSED, 2D-MTS and 2D-MERR fracture models [46], the theoretical fracture toughness ratio K_{IIc}/K_{Ic} can be obtained under the given Poisson coefficient ν, as shown in Fig. 4.9a. Obviously, the fracture toughness ratios K_{IIc}/K_{Ic} predicted by the 2D-MTS and 2D-MERR fracture models are 0.87 and 0.63, respectively, indicating that the variation of the Poisson coefficient ν is ineffective. The fracture toughness ratio K_{IIc}/K_{Ic} predicted by the 2D-MSED fracture model decreases with the increasing v values. It is generally acknowledged that shear strength (i.e., mode II or mode III fracture) is significantly greater than tension strength (i.e., mode I fracture). However, the maximum fracture toughness ratio K_{IIc}/K_{Ic} predicted by these three established fracture models is only 1.2. A reasonable explanation for the disparity of K_{IIc}/K_{Ic} is that the aforementioned fracture models are derived and established based on the tensile mathematical frameworks. In other words, pure mode-I fracture toughness K_{Ic} plays an important role in these established fracture models. Consequently, the predicted mode-II fracture toughness depends on the inherent material index (i.e., pure mode-I fracture toughness K_{Ic}). In addition, these fracture models mainly rely on the Poisson coefficient v to predict the fracture toughness ratio K_{IIc}/K_{Ic}. Such calculations fail to fully characterize and reflect the true fracture mechanisms of tested materials. Taking the plane stress case as an example, Fig. 4.9b shows the theoretical fracture toughness ratio K_{IIc}/K_{Ic} predicted by the 2D-MADSED (modified two-dimensional distortional strain energy density) criterion. It can be seen from Fig. 4.9b that the fracture toughness ratio K_{IIc}/K_{Ic} predicted by the current fracture criterion shows an increasing tendency with the increasing critical parameter $\eta = r_{IIc}/r_{Ic}$. With the decrease of the critical parameter η, the fracture mechanism gradually changes from shear-dominated failure to tension-dominated failure. This phenomenon implies that the 2D-MADSED criterion can not only provide accurate evaluation for mode-II loading conditions (i.e., tension-dominated fracture), but also provide successful prediction for mode-II fracture conditions (i.e., shear-dominated fracture).

Taking the plane stress case as an example, Fig. 4.10 shows the influence of the critical parameter η (i.e., r_{IIc}/r_{Ic}) on the combined-mode I/II fracture toughness. Obviously, the predicted combined-mode I/II fracture toughness gradually increases with the increasing critical parameter η. This shows that the critical elastic–plastic zone radius can significantly affect the combined-mode I/II fracture resistance of materials, namely the larger the critical elastic–plastic zone radius, the greater the fracture resistance. In addition, the macroscopic fracture begins to appear at the

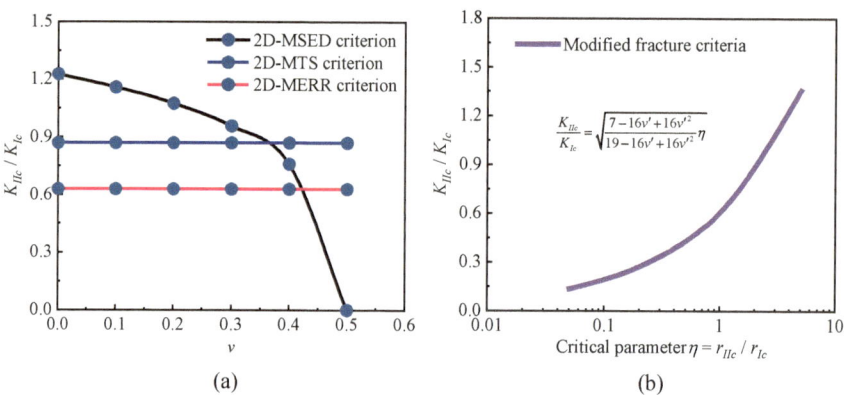

Fig. 4.9 Fracture toughness ratio K_{IIc}/K_{Ic} predicted by fracture criteria

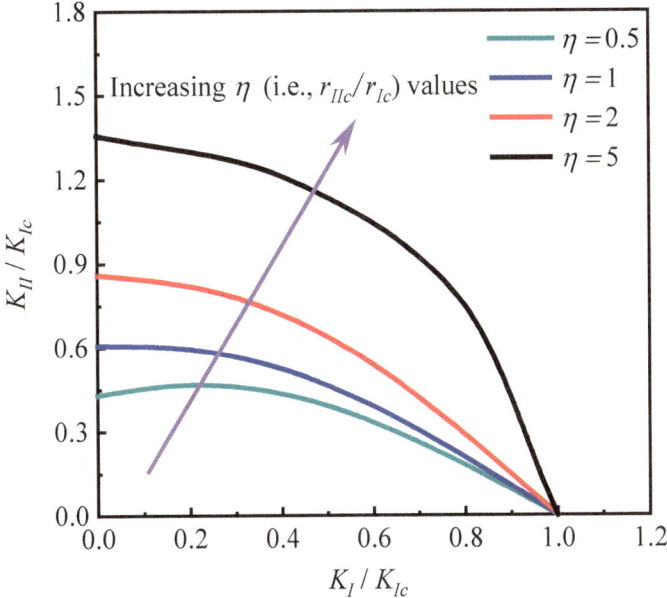

Fig. 4.10 Effect of the critical parameter η on mixed-mode I/II fracture toughness

distance from the crack tip to the critical elastic–plastic boundary. Actually, the critical elastic–plastic zone radius increases, requiring more energy dissipated to create the elastic–plastic zone, which leads to greater resistance to cracking propagation. Therefore, the fracture resistance can be improved by the increasing critical elastic–plastic zone radius. Consequently, the modified fracture criterion can theoretically analyze the fracture mechanisms of materials under different mixed-mode I/II loadings, which further proves the rationality and applicability of the current fracture criterion. The modified average distortional strain energy density criterion can provide a certain theoretical guidance for the analyses of crack propagation and elastic–plastic behaviors.

Different combined-mode I/II fracture test methods are used to further validate the modified average distortional strain energy density criterion, as shown in Fig. 4.11, the tested data in Fig. 4.11a is obtained by the CCBD (centrally cracked Brazilian disk) test method [47], the tested data in Fig. 4.11b is acquired from the SBB (short beam bending) test method [48], the tested data in Fig. 4.11c is determined from the SCB test method [49], and the tested data in Fig. 4.11d is obtained by the ECTB test method [50]. It can be observed from Fig. 4.11 that the combined-mode I/II fracture resistance of materials significantly depends on the loading and geometry configurations of test specimens. Compared with the traditional and common 2D-MTS fracture criterion, the modified average distortional strain energy density criterion can provide accurate evaluation and prediction for combined-mode I/II fracture toughness under different fracture testing methods. Figure 4.12 shows the contrast

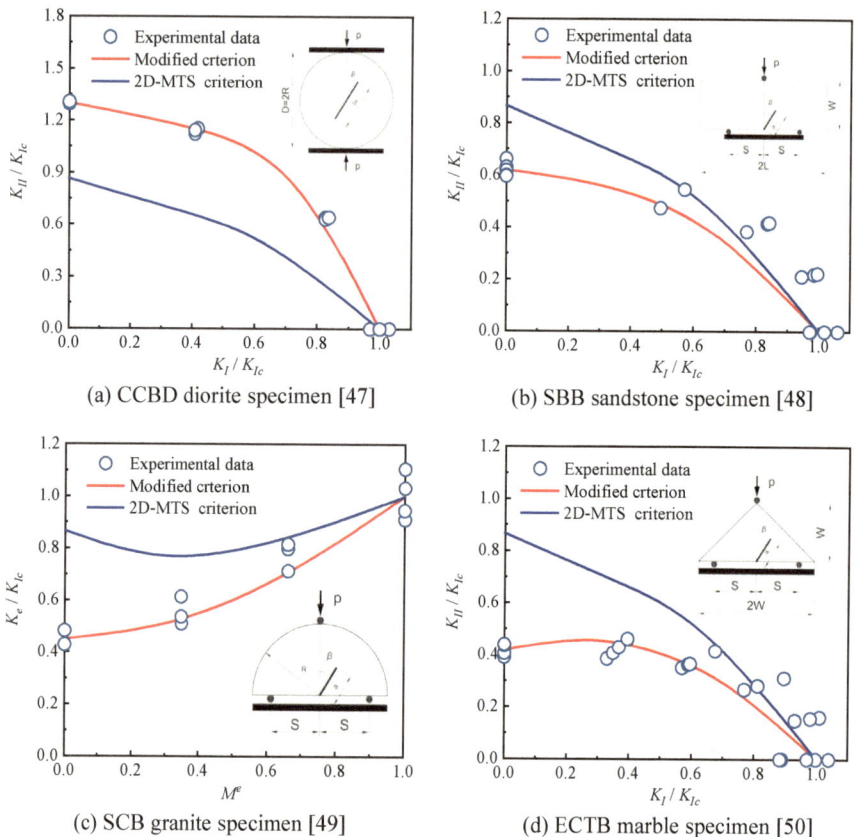

Fig. 4.11 Verification of the improved criterion using different mixed-mode I/II testing methods

between the normalized effective plastic zone radius and the normalized traditional plastic zone radius. Transitioning from pure mode-I loading to pure mode-II loading, the traditional plastic zone radiuses are constant. However, the effective plastic zone radiuses are variable and depend on the specimen geometry and loading form of materials, which is more consistent with the microscopic deformation mechanisms of materials under different stress conditions.

4.4   A Novel Improved Semi-Circular Bend Specimen

Concrete [51], rocks [52, 53], gypsum, asphalt [54], and other materials exhibit brittle behavior with brittle fracture being the main failure mode for such materials [55]. In this case, Linear Elastic Fracture Mechanics (LEFM) provides a good framework

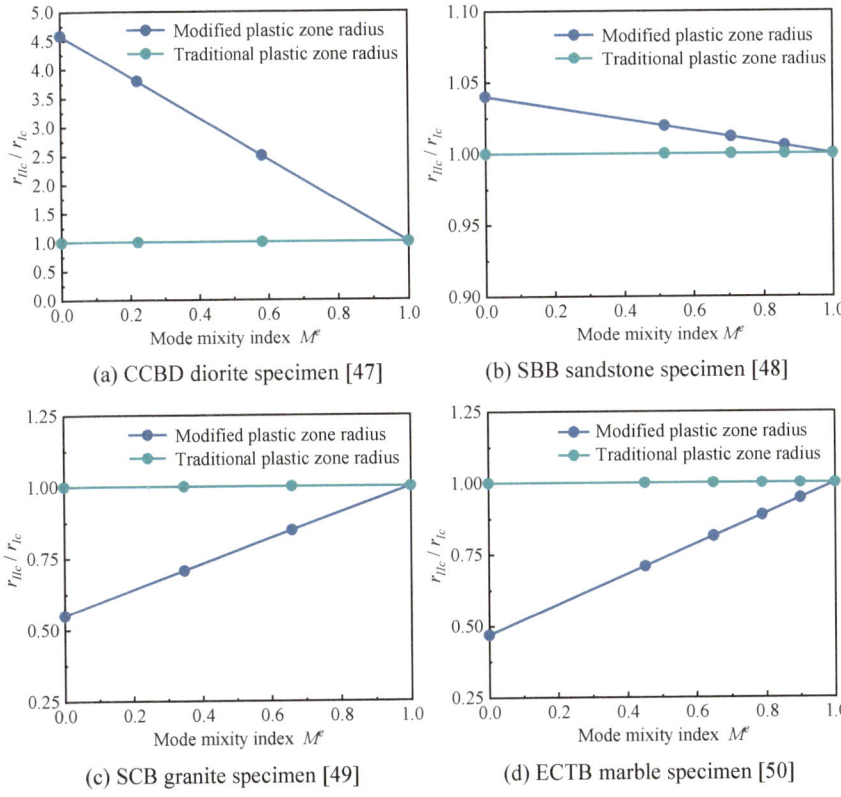

Fig. 4.12 Comparison between the improved plastic zone radius and the traditional plastic zone radius

for evaluating fracture parameters where fracture toughness is the primary parameter describing resistance to crack propagation.

The study of fracture toughness necessitates experimental research using appropriate laboratory-scale specimens. In this regard, numerous testing devices have been proposed by previous researchers over the past few decades. Table 4.2 presents several commonly used experimental methods for simulating fracture toughness of brittle materials.

However, in practical engineering, special structures are often encountered. In order to study the fracture toughness of complex structures, many researchers have proposed new testing devices. Wu [82] proposed a novel test device for investigating the fracture toughness of structures during tunnel excavation. Karimi [93] introduced an improved ring specimen to expand its applicability range. Gope [94] developed an enhanced semi-circular bending experimental method to investigate the fracture toughness of epoxy-based bio-composite materials. Trajkovi [95] presented a innovative experimental apparatus for studying the fracture toughness of pipeline materials under tensile loading.

Table 4.2 Common testing methods

Specimen Name	Shape	Reference
Brazilian disc (BD)	Disc	Aliha [56], Akbardoost [57], and He [58]
Asymmetrical edge-notched disc bend (A-ENDB) specimen	Disc	Mousavi [59] and Bahmani [60]
Single edge notched bend (SENB) specimen	Beam	Marsavina [61], Luo [62], and Linul [63]
Asymmetric four-point bending (AFPB) specimen	Beam	Apostol [64], Linul [65], and Yoshihara [66]
Short bend beam (SBB) specimen	Beam	Aliha [67, 68], Mousavi [69], and Saed [70]
Notched semi-circular specimen in bending (SCB)	Semi-circular	Ayatollahi [71], Kuruppu [72], Bahrami [73], Mirsayar [74], Hou [75], Pirmohammad [76], Bahrami [77], and Karamzadeh [78]
Semi-circular bend (SCB) specimen with asymmetrical supports	Semi-circular	Ayatollahi [79], Nejati [80], and Torabi [81]
Single edge crack (SEC) specimen	Rectangular plate	Marsavina [82], and Hammond [83], and Saxena [84]
Center cracked ring-shape specimen	Ring	Zhou [85], Chen [86], Eftekhari [87], Amrollahi [88], Akbardoost [89], Aliha [90], and Zhou [91]

The above devices did not take into account how excavation affects the fracture toughness of structures. To examine how excavated holes affect structural fracture toughness, this paper introduces a novel non-matching semi-circular bending experimental device that builds upon the classical SCB test. This modified device includes a semi-circular hole and applies asymmetric structural constraints to accurately assess its impact. Further details regarding this analysis will be provided in the following section.

4.4.1 Improved SCB Device

The improved SCB specimen is depicted in Fig. 4.13, illustrating its geometric shape and loading conditions. The specimen consists of a semicircular ring with an outer radius R, inner radius R_i, and thickness t. a represents the distance from the center to the crack tip, while S_1 and S_2 denote the left and right loading points respectively, with loading direction parallel to the crack direction. By adjusting the positions of two bottom supports (S_1 and S_2) during experimentation, it becomes easy to alter the mode mixity in ISCB specimens. When bottom loads are applied directly on the crack line (i.e., when $S_1 = S_2$), pure mode I influence is observed in the specimen.

Fig. 4.13 Diagram of the ISCB device

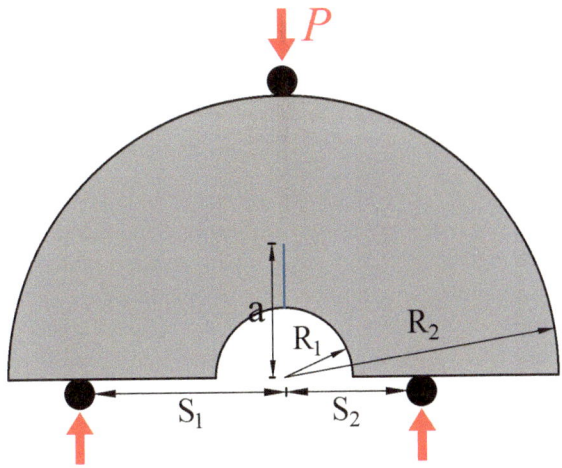

However, for asymmetric loading (i.e., when $S_1 \neq S_2$), control over contributions from both mode I and mode II can be achieved by selecting appropriate values for both S_1 and S_2. Therefore, different mode mixities can be obtained in this proposed specimen design. This simple structure allows for analyzing how voids in materials such as rocks, concrete, asphalt, polymers impact structural fracture toughness.

Compared to the SCB specimen, the ISCB specimen has several advantages. It is easier to create a longer fracture path and increase stress concentration at the crack tip in the rock core sample. The semi-circular hole also enhances beneficial stress concentration at the crack tip and offers significant advantages in observing mode-I dynamic fracture processes. Additionally, pre-cracking is convenient without requiring loading fixtures. The proposed specimen type allows for an extended fracture zone to fully capture the entire fracture process, and the stress pattern of the specimen transitions from compression to bending.

To investigate the mixed-mode fracture of brittle materials using this configuration, it is necessary to calculate the stress intensity factors for mode I and mode II (K_I and K_{II}) at different loading positions and crack lengths. This study utilizes the finite element method to determine K_I and K_{II} in ISCB specimens. More details regarding these calculations will be provided in the next section.

4.4.2 Finite Element Analysis

The stress intensity factors K_I and K_{II} for improving the ASCB specimen are defined as functions of crack length (a) and load support positions, denoted by S_1 and S_2. Referring to the calculation formula of SCB, the stress intensity factor for ISCB can be expressed as follows.

$$Y_I = \frac{K_I 2(R - R_i)t}{P\sqrt{\pi(a - R_i)}} \qquad (4.12)$$

$$Y_{II} = \frac{K_{II} 2(R - R_i)t}{P\sqrt{\pi(a - R_i)}} \qquad (4.13)$$

where t is the thickness of the specimen and K is the stress intensity factor.

The ISCB specimen was analyzed using different finite element models with the ABAQUS code. Figure 4.14 displays the typical mesh pattern generated for simulating the specimen, which considered geometric shapes and loading conditions of R = 60 mm, t = 6 mm, P = 1000 N, and varying crack length values. S_1 was fixed at a value of 40 mm while S_2 varied from zero to 40 mm to alter the mode mixity state. The elastic material properties were taken into account in the finite element model with E = 2970 MPa and v = 0.35. Each model utilized a total of 1042 eight-node plane stress elements, including singular elements in the first ring around the crack tip to generate square root singularity in stress/strain fields. The software directly obtained stress intensity factors using ABAQUS's built-in J-integral based method.

Figures 4.15 and 4.16 illustrate the values of Y_1 and Y_2 for different crack lengths a and R_i under S = 40 mm. Figures 4.17 and 4.18 illustrate the values of Y_1 and Y_2 for different crack lengths a and R_i under S = 48 mm.

From these graphs, it can be observed that under symmetric loading conditions (i.e., $S_1 = S_2$), Y_{II} equals zero, indicating that the specimen is subjected to pure Mode I loading. By changing the position of S_2, a Mode II component also appears in the ISCB specimen. As shown in Figs. 4.15 and 4.16, by moving S_2 along the crack plane, the geometric factor Y_I for mode I decreases while the geometric factor Y_{II} for Mode II increases. With an increase in Ri, both Y_I and Y_{II} decrease; however, the decrease in Y_{II} is significantly greater than that of Y_I, suggesting a larger influence of inner radius R_i on Y_{II}.

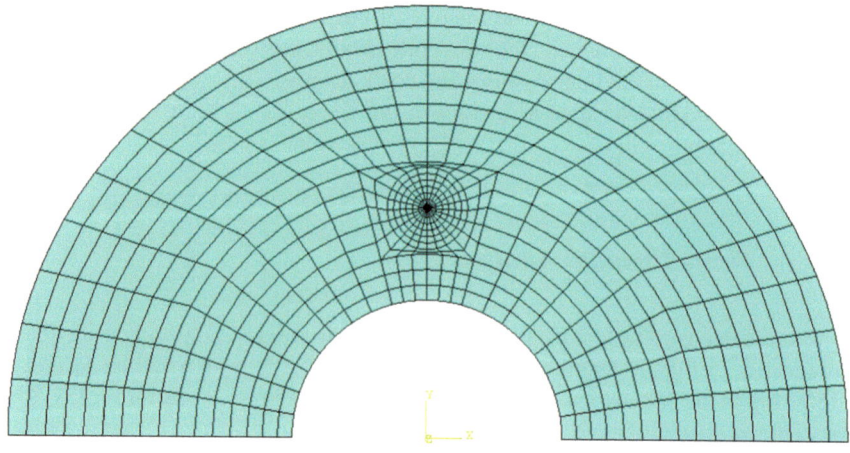

Fig. 4.14 Finite element model of the ISCB device

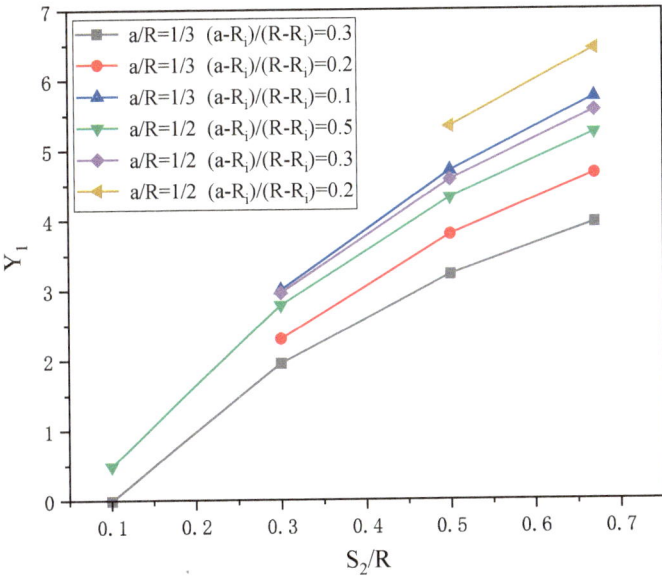

Fig. 4.15 Different Y_I values at $S = 40$ mm

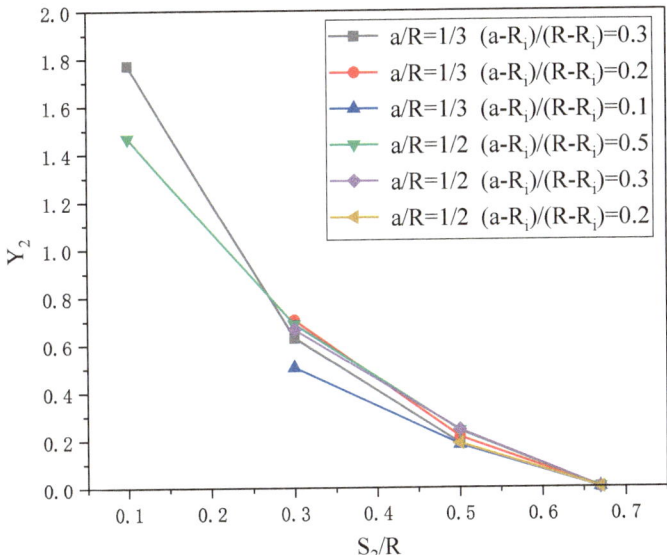

Fig. 4.16 Different Y_{II} values at $S = 40$ mm

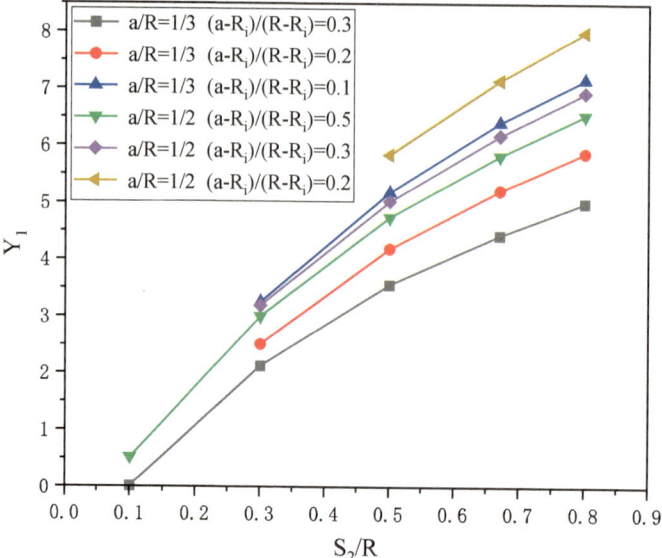

Fig. 4.17 Different Y_I values at S = 48 mm

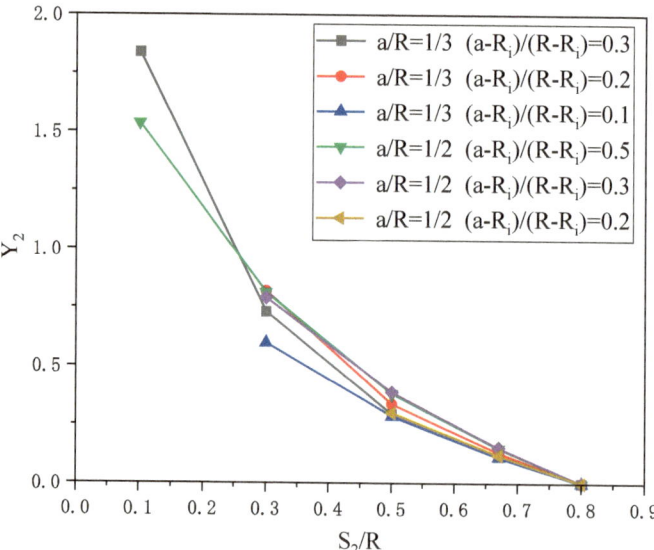

Fig. 4.18 Different Y_{II} values at S = 48 mm

4.5 Conclusion

The SCB test method is applied to investigate the pure mode-I, combined-mode I + II, and pure mode-II fracture toughness. We consider the effective critical plastic zone radius r_{ec} instead of the conventional invariable radius r_{Ic} to improve the traditional average distortional strain energy density criterion. The leading conclusions are outlined as below:

(1) For the tested SCB sandstone, the fracture load (i.e., peak load) tends to increase when transforming from pure mode-I loading to pure mode-II loading, whereas the fracture toughness tends to decrease.

(2) The reasonable selection of the fracture progress zone size rc plays an important role in improving the prediction accuracy of the GMTS fracture model. The modified average distortional strain energy density criterion can provide accurate evaluation and prediction for combined-mode I/II fracture toughness under different fracture testing methods.

(3) For mixed mode I/II fracture, a new experimental configuration called the improved semi-circular bend (ISCB) specimen is proposed to test the fracture behaviors of brittle materials. The main advantage of ISCB specimens lies in their simple geometric shape and loading setup, which facilitate crack initiation within the specimen and allow for a combination of complete mode I and mode II conditions. The excavation of holes significantly affects the structural response, as increasing inner diameters leads to reduced fracture toughness of cracks and has a noticeable influence on K_{II}.

References

1. Ren L, Xie LZ, Xie HP, Ai T, He B (2016) Mixed-mode fracture behavior and related surface topography feature of a typical sandstone. Rock Mech Rock Eng 49:3137–3153
2. Zhao Y, Zheng K, Wang CL, Bi J, Zhang H (2022) Investigation on model-I fracture toughness of sandstone with the structure of typical bedding inclination angles subjected to three-point bending. Theor Appl Fract Mech 119:103327
3. Liu J, Li Y, Qiao L (2022) Analytical solutions of stress intensity factors for a centrally cracked Brazilian disc considering tangential friction effects. Rock Mech Rock Eng
4. Awaji H, Sato S (1978) Combined mode fracture toughness measurement by the disk test. J Eng Mater Technol 100:175–182
5. Atkinson C, Smelser RE, Sanchez J (1982) Combined mode fracture via the cracked Brazilian disk test. Int J Fract 18(4):279–291
6. Fett T (2011) Stress intensity factors and *T*-stress for internally cracked circular disks under various boundary conditions. Eng Fract Mech 68:119–136
7. Dong S, Wang Y, Xia Y (2004) Stress intensity factors for central cracked circular disk subjected to compression. Eng Fract Mech 71:1135–1148
8. Hua W, Dong SM, Li YF, Xu JG, Wang QY (2015) The influence of cyclic wetting and drying on the fracture toughness of sandstone. Int J Rock Mech Min 78:331–335
9. Hua W, Dong SM, Li YF, Wang QY (2016) Effect of cyclic wetting and drying on the pure mode II fracture toughness of sandstone. Eng Fract Mech 153:143–150

10. Hua W, Dong SM, Peng F, Li KY, Wang QY (2017) Experimental investigation on the effect of wetting-drying cycles on mixed mode fracture toughness of sandstone. Int J Rock Mech Min 93:242–249

11. Hua W, Dong SM, Fan Y, Pan X, Wang QY (2017) Investigation on the correlation of mode II fracture toughness of sandstone with tensile strength. Eng Fract Mech 184:249–258

12. Aliha MRM, Ayatollahi MR, Akbardoost J (2012) Typical upper bound–lower bound mixed mode fracture resistance envelopes for rock material. Rock Mech Rock Eng 45:65–74

13. Ayatollahi MR, Aliha MRM (2011) Fracture analysis of some ceramics under mixed mode loading. J Am Ceram Soc 94(2):561–569

14. Nejati M, Bahrami B, Ayatollahi MR, Driesner T (2021) , Thomas Driesner, On the anisotropy of shear fracture toughness in rocks. Theor Appl Fract Mech 113:102946

15. Dou FK, Wang JG, Zhang XX, Wang HM (2019) Effect of joint parameters on fracturing behavior of shale in notched three-point-bending test based on discrete element model. Eng Fract Mech 205:40–56

16. Liu J, Yao K, Xue Y, Zhang XX, Chong ZH, Liang X (2019) Study on fracture behavior of bedded shale in three-point-bending test based on hybrid phase-field modelling. Theor Appl Fract Mech 104:102382

17. Li YJ, Wang ST, Zheng LG, Zhao SK, Zuo JP (2021) Evaluation of the fracture mechanisms and criteria of bedding shale based on three-point bending experiment. Eng Fract Mech 255:107913

18. Zhao YX, Sun Z, Gao YR, Wang XL, Song HH (2022) Influence of bedding planes on fracture characteristics of coal under mode II loading. Theor Appl Fract Mech 117:103131

19. Sun DL, Rao QH, Wang SY, Shen QQ, Yi W (2021) Shear fracture (Mode II) toughness measurement of anisotropic rock. Theor Appl Fract Mech 115:103043

20. Shi XS, Zhao YX, Danesh NN, Zhang X, Tang TW (2022) Role of bedding plane in the relationship between Mode-I fracture toughness and tensile strength of shale. B Eng Geol Environ 81:81

21. Wang H, Li Y, Cao SG, Fantuzzi N, Pan RK, Tian MY, Liu YB, Yang HY (2020) Fracture toughness analysis of HCCD specimens of Longmaxi shale subjected to mixed mode I-II loading. Eng Fract Mech 239:107299

22. Hou P, Su SJ, Liang X, Gao F, Cai CZ, Yang YG, Zhang ZZ (2021) Effect of liquid nitrogen freeze–thaw cycle on fracture toughness and energy release rate of saturated sandstone. Eng Fract Mech 258:108066

23. Mirsayar MM, Razmi A, Aliha MRM, Berto F (2018) EMTSN criterion for evaluating mixed mode I/II crack propagation in rock materials. Eng Fract Mech 190:186–197

24. Ayatollahi MR, Sistaninia M (2011) Mode II fracture study of rocks using Brazilian disk specimens. Int J Rock Mech Min 48:819–826

25. Cao RH, Wang CS, Yao RB, Hu T, Lei DX, Lin H, Zhao YL (2020) Effects of cyclic freeze-thaw treatments on the fracture characteristics of sandstone under different fracture modes: laboratory testing. Theor Appl Fract Mech 109:102738

26. Kang P, Hong L, Hong, Fazhi Y, Quanle Z, Xiao S, Zhaopeng L (2020) Effects of temperature on mechanical properties of granite under different fracture modes. Eng Fract Mech 226:106838

27. Liu J, Qiao L, Li Y, Li QW, Fan DJ (2022) Experimental study on the quasi-static loading rate dependency of mixed-mode I/II fractures for marble rocks. Theor Appl Fract Mech 121:103431

28. Aminzadeh A, Bahrami B (2022) Majid Reza Ayatollahi, Morteza Nejati, On the role of fracture process zone size in specifying fracturing mechanism under dominant mode II loading. Theor Appl Fract Mech 117:103150

29. Khan K, Al-Shayea NA (2000) Effect of specimen geometry and testing method on mixed mode I-II fracture toughness of a limestone rock from Saudi Arabia. Rock Mech Rock Eng 33(3):179–206

30. Lim IL, Johnston IW, Choi SK, Boland JN (1994) Fracture testing of a soft rock with semi-circular specimens under three-point bending. Part 2: mixed-mode 31(3):185–197

31. Ayatollahi MR, Aliha MRM (2007) Wide range data for crack tip parameters in two disc-type specimens under mixed mode loading. Comp Mater Sci 38:660–670

32. Ayatollahi MR, Akbardoost J (2013) Size effects in mode II brittle fracture of rocks. Eng Fract Mech 112–113:165–180

33. Xie Q, Liu XL, Li SX, Du K, Gong FQ, Li XB (2022) Prediction of mode I fracture toughness of shale specimens by different fracture theories considering size effect. Rock Mech Rock Eng 55:7289–7306

34. Zhao GL, Yao W, Li X, Xu Y, Xia KW, Chen R (2022) Influence of notch geometry on the rock fracture toughness measurement using the ISRM suggested semi-circular bend (SCB) Method. Rock Mech Rock Eng 55:2239–2253

35. Bahrami B, Ayatollahi MR, Sedighi I, Yahya MY (2019) An insight into mode II fracture toughness testing using SCB specimen. Fatigue Fract Eng Mater Struct 42:1991–1999

36. Pirmohammad S, Abdi M, Ayatollahi MR (2021) Mode II fracture tests on asphalt concrete at different temperatures using semi-circular bend specimen loaded by various types of supports. Theoret Appl Fract Mech 116:103089

37. Obara YZ, Nakamura K, Yoshioka S, Sainoki A, Kasai A (2019) Crack front geometry and stress intensity factor of semi-circular bend specimens with straight through and chevron notches. Rock Mech Rock Eng

38. Aliha MRM, Mahdavi E, Ayatollahi MR (2016) The influence of specimen type on tensile fracture toughness of rock materials. Pure Appl Geophys

39. Aliha MRM, Mahdavi E, Ayatollahi MR () Statistical analysis of rock fracture toughness data obtained from different chevron notched and straight cracked mode I specimens, Rock Mech, Rock Eng (2018)

40. Ren L, Zhu Z, Wang M, Zheng T, Ai T (2014) Mixed-mode elastic-plastic fractures: improved r-criterion. J Eng Mech 140(6):04014033

41. Kun Z, Chaolin W, Zhao Y, Jing B (2023) Theoretical and experimental researches on fracture toughness for bedded shale using the centrally cracked Brazilian disk method with acoustic emission monitoring. Theoret Appl Fract Mech 124:103784

42. Dehghani B, Faramarzi L (2019) Experimental investigations of fracture toughness and crack initiation in marble under different freezing and thermal cyclic loading. Constr Build Mater 220:340–352

43. Kun Z, Zhao Y, Chaolin W, Jing B (2024) Influence of distinct testing methods on the mode-I fracture toughness of Longmaxi shale. Theoret Appl Fract Mech 129:104213

44. Williams ML (1957) On the stress distribution at the base of a stationary crack. J Appl Mech 24:109–114

45. Wei Y (2012) An extended strain energy density failure criterion by differentiating volumetric and distortional deformation. Int J Solids Struct 49:1117–1126

46. Wang J, Ren L, Xie LZ, Xie HP, Ai T (2016) Maximum mean principal stress criterion for three-dimensional brittle fracture. Int J Solids Struct 102–103:142–154

47. Zhu Q, Li D, Li X, Han Z, Ma J (2023) Mixed mode fracture parameters and fracture characteristics of diorite using cracked straight through Brazilian disc specimen. Theoret Appl Fract Mech 123:103682

48. Luo Y, Ren L, Xie LZ, Ai T, He B (2017) Fracture behavior investigation of a typical sandstone under mixed-mode I/II loading using the notched deep beam bending method. Rock Mech Rock Eng 50:1987–2005

49. Feng G, Zhu C, Wang XC, Tang SB (2023) Thermal effects on prediction accuracy of dense granite mechanical behaviors using modified maximum tangential stress criterion. J Rock Mech Geotech Eng 15:1734–1748

50. Aliha MRM, Hosseinpour GhR, Ayatollahi MR (2013) Application of cracked triangular specimen subjected to three-point bending for investigating fracture behavior of rock materials. Rock Mech Rock Eng 46:1023–1034

51. Abolhasani A, et al. (2022) A comprehensive evaluation of fracture toughness, fracture energy, flexural strength and microstructure of calcium aluminate cement concrete exposed to high temperatures. Eng Fract Mech 261

52. Tutluoglu L, Batan CK, Aliha MRM (2022) Tensile mode fracture toughness experiments on andesite rock using disc and semi-disc bend geometries with varying loading spans. Theoret Appl Fract Mech 119

53. Afrasiabian B, Eftekhari M (2022) Prediction of mode I fracture toughness of rock using linear multiple regression and gene expression programming. J Rock Mech Geotech Eng 14(5):1421–1432

54. Wen X et al. (2022) Relationship between low-temperature KIc and KIIc values of bitumen with different performance grades and comparison with naturally solid materials. Theoret Appl Fract Mech 117

55. Zhang N et al (2022) Assessment of fiber factor for the fracture toughness of polyethylene fiber reinforced geopolymer. Constr Build Mater 319:126130

56. Aliha MRM, Mahdavi E, Ayatollahi MR (2017) The influence of specimen type on tensile fracture toughness of rock materials. Pure Appl Geophys 174(3):1237–1253

57. Akbardoost J et al (2014) Size-dependent fracture behavior of Guiting limestone under mixed mode loading. Int J Rock Mech Mining Sci 71:369–380

58. He J et al. (2021) Contribution of interface fracture mechanism on fracture propagation trajectory of heterogeneous asphalt composites. Appl Sci-Basel 11(7)

59. Mousavi SR, et al. (2021) Effect of waste glass and curing aging on fracture toughness of self-compacting mortars using ENDB specimen. Construct Build Mater 282

60. Bahmani A et al (2020) An extended edge-notched disc bend (ENDB) specimen for mixed-mode I+II fracture assessments. Int J Solids Struct 193–194:239–250

61. Marsavina L et al (2014) Refinements on fracture toughness of PUR foams. Eng Fract Mech 129:54–66

62. Luo Y et al (2017) Fracture behavior investigation of a typical sandstone under mixed-mode I/II loading using the notched deep beam bending method. Rock Mech Rock Eng 50(8):1987–2005

63. Linul E, et al. (2020) Static and dynamic mode I fracture toughness of rigid PUR foams under room and cryogenic temperatures. Eng Fract Mech 225

64. Apostol DA et al (2016) Crack length influence on stress intensity factors for the asymmetric four-point bending testing of a polyurethane foam. Mater Plast 53(2):280–282

65. Linul E, Marsavina L, Stoia DI (2020) Mode I and II fracture toughness investigation of Laser-Sintered Polyamide. Theoret Appl Fract Mech 106

66. Yoshihara H, Maruta M (2021) Mode II critical stress intensity factor of solid wood obtained from the asymmetric four-point bend fracture test using groove-free and side-grooved samples. Eng Fract Mech 258

67. Aliha MRM, Samareh-Mousavi SS, Mirsayar MM (2021) Loading rate effect on mixed mode I/II brittle fracture behavior of PMMA using inclined cracked SBB specimen. Int J Solids Struct 232

68. Aliha MRM, Karimi HR, Ghoreishi SMN (2022) Design and validation of simple bend beam specimen for covering the full range of I+II fracture modes. Eur J Mech A Solids 91:104425

69. Mousavi SS, Aliha MRM, Imani DM (2020) On the use of edge cracked short bend beam specimen for PMMA fracture toughness testing under mixed-mode I/II. Polym Testing 81

70. Saed SA et al. (2022) Full range I/II fracture behavior of asphalt mixtures containing RAP and rejuvenating agent using two different 3-point bend type configurations. Constr Build Mater 314

71. Ayatollahi MR, Aliha MRM, Hassani MM (2006) Mixed mode brittle fracture in PMMA—An experimental study using SCB specimens. Mater Sci Eng, A 417(1–2):348–356

72. Kuruppu MD, Chong KP (2012) Fracture toughness testing of brittle materials using semi-circular bend (SCB) specimen. Eng Fract Mech 91:133–150

73. Bahrami B et al (2019) An insight into mode II fracture toughness testing using SCB specimen. Fatigue Fract Eng Mater Struct 42(9):1991–1999

74. Mirsayar MM, Razmi A, Berto F (2018) Tangential strain-based criteria for mixed-mode I/II fracture toughness of cement concrete. Fatigue Fract Eng Mater Struct 41(1):129–137

75. Hou C et al (2019) A generalized maximum energy release rate criterion for mixed mode fracture analysis of brittle and quasi-brittle materials. Theoret Appl Fract Mech 100:78–85

76. Pirmohammad S, Abdi M, Ayatollahi MR (2021) Mode II fracture tests on asphalt concrete at different temperatures using semi-circular bend specimen loaded by various types of supports. Theoret Appl Fract Mech 116

77. Bahrami B et al (2019) An insight into mode II fracture toughness testing using SCB specimen. Fatigue Fract Eng Mater Struct 42(9):1991–1999
78. Karamzadeh NS, Aliha MRM, Karimi HR (2022) Investigation of the effect of components on tensile strength and mode-I fracture toughness of polymer concrete. Arab J Geosci 15(13)
79. Ayatollahi MR, Aliha MRM, Saghafi H (2011) An improved semi-circular bend specimen for investigating mixed mode brittle fracture. Eng Fract Mech 78(1):110–123
80. Nejati M, Ghouli S, Ayatollahi MR (2020) Crack tip asymptotic field and K-dominant region for anisotropic semicircular bend specimen. Theoret Appl Fract Mech 109
81. Torabi E et al. (2021) Mixed mode fracture behavior of short-particle engineered wood. Theoret Appl Fract Mech 115
82. Marsavina L et al (2015) Shear and mode II fracture of PUR foams. Eng Fail Anal 58:465–476
83. Hammond MJ, Fawaz SA (2016) Stress intensity factors of various size single edge-cracked tension specimens: A review and new solutions. Eng Fract Mech 153:25–34
84. Saxena A et al (2017) On single-edge-crack tension specimens for tension-compression fatigue crack growth testing. Eng Fract Mech 176:343–350
85. Zhou X et al. (2019) Comprehensive study on the crack tip parameters of two types of disc specimens under combined confining pressure and diametric concentrated forces. Theoret Appl Fract Mech 103
86. Chen CH, Chen CS, Wu JH (2008) Fracture toughness analysis on cracked ring disks of anisotropic rock. Rock Mech Rock Eng 41(4):539–562
87. Eftekhari M et al (2015) Mechanism of fracture in macro- and micro-scales in hollow centre cracked disc specimen. J Central South Univ 22(11):4426–4433
88. Amrollahi H, Baghbanan A, Hashemolhosseini H (2011) Measuring fracture toughness of crystalline marbles under modes I and II and mixed mode I-II loading conditions using CCNBD and HCCD specimens. Int J Rock Mech Mining Sci 48(7):1123–1134
89. Akbardoost J, Ghadirian HR, Sangsefidi M (2017) Calculation of the crack tip parameters in the holed-cracked flattened Brazilian disk (HCFBD) specimens under wide range of mixed mode I/II loading. Fatigue Fract Eng Mater Struct 40(9):1416–1427
90. Aliha MRM, Ayatollahi MR, Pakzad R (2008) Brittle fracture analysis using a ring-shape specimen containing two angled cracks. Int J Fract 153(1):63–68
91. Zhou X, Wang L, Shou Y (2020) Understanding the fracture mechanism of ring Brazilian disc specimens by the phase field method. Int J Fract 226(1):17–43
92. Wu L et al. (2022) Rock dynamic fracture of a novel semi-circular-disk specimen. Int J Rock Mech Mining Sci 152
93. Karimi HR, et al. (2022) A comprehensive study on ring shape specimens under compressive and tensile loadings for covering the full range of I plus II fracture modes of gypsum material. Int J Rock Mech Mining Sci 160
94. Gope PC (2018) Maximum tangential stress coupled with probabilistic aspect of fracture toughness of hybrid bio-composite. Eng Sci Technol-An Int J-JESTECH 21(2):201–214
95. Trajkovic I, et al. (2023) Selective laser sintered Pipe Ring Notched Tension specimens for examination of fracture properties of pipeline materials. Eng Fract Mech 292

Chapter 5
Ture Mode-II Fracture

The HF (hydraulic fracturing) technique has received widespread recognition and applications in the extraction of shale oil/gas, coalbed methane, and geothermal system [1–3]. The primary objective of HF is to optimize the fracture networks within rock reservoirs, thereby enhancing productivity and recovery [4–6]. As a significant mechanical indicator, the SIF (stress intensity factor) can weigh the stress and deformation fields at the crack tip adjacency, and the onset of fracturing is imminent when the SIF approaches its critical level (i.e., fracture toughness K_c) [7–17]. According to classical definitions of fracture modes, the mode-I (tension mode) fracturing involves the symmetrical opening of cracking sides, while the mode-II (shear mode) fracturing involves the self-similar sliding of cracking sides [18–21]. At present, numerous test techniques and corresponding specimen configurations have been established and modified for determining the fracture toughness of engineering materials [22–28]. The combined mode I + II fracturing resistance has been emphasized using the popular experiment fixtures, including the CCBD (centrally cracked Brazilian disk), SCB (semicircular bending), ECTB (edge-cracked triangle bending), ENRBB (edge-notched rectangular beam bending), and CCRD-DC (centrally cracked ring disk in diametral compression) specimens [29–39]. Under specific geometry and loading states, the aforementioned test specimens are capable of measuring the pure mode-II fracturing toughness K_{IIc}. However, the mode-II cracking under pure mode-II loading deviates at an angle about 70° relative to original crack front and does not develop in a self-planar and shear-driven manner, resulting in measured uncertainty regarding the true mode-II fracturing. Moreover, the formation of the kinked fracture trajectory is driven by the tension stress. This implies that the mode-II testing results of these suggested specimens are based on the mode-II loading rather than the true mode-II fracturing [40–44], as displayed in Fig. 5.1. Hence, the mode-II fracturing toughness associated with these tension-based tests should be recognized as apparent K_{IIC} rather than true K_{IIC}, is not considered an independently inherent material property, and depends on the mode-I fracturing toughness K_{Ic} [40].

© The Author(s) 2024
Y. Zhao et al., *Rock Fracture Mechanics and Fracture Criteria*,
https://doi.org/10.1007/978-981-97-5822-7_5

Fig. 5.1 Comparison
between apparent mode II
fracture and true mode II
fracture

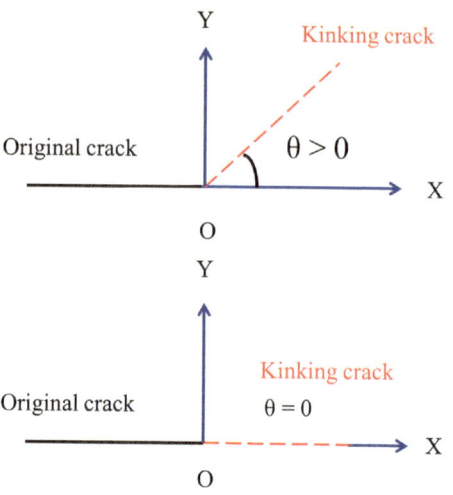

Since deep shale gas reservoirs with natural cracks are mostly subjected to compressive stresses produced by tectonic stress and rock weight, the rupture of rock bridges between two adjoining discontinuities within rocks easily develops along the maximum shearing stress orientation in a sliding manner (true mode II) [45–54]. To prevent the occurrence of curvilinear cracking trajectory, the qualified mechanical system should be required for the evaluation of true K_{IIc}. Only three test specimens of true K_{IIc}, unlike the measurement of K_{Ic}, are applied because of their capability to accomplish both the mode-II loading and shear-induced mode-II fracturing. Available test specimens include (1) the PTS-CP (punch-through shearing with confined pressure) specimen proposed by [55], (2) the DEND-DC (double-edge notched disk in diametral compression) specimen developed by [42], and (3) the SB-SENC or SB-DENC (shear-box with single-edge or double-edge notched cube) specimen established by [56]. The DEND-DC test specimen is a successful method for determining true K_{IIc}, however, it needs to be further explored on the influences of the size and stiffness of flexible jaw. While the SB-SENC or SB-DENC test specimen requires complicated experiment techniques (e.g., cubic sample processing, fixture alignment precision, and uncommon loading equipment). In addition, the compressive and shearing stresses in DEND-DC and SB-SENC or SB-DENC tests cannot be applied independently. Due to the limitations of DEND-DC and SB specimens, the straightforward PTS-CP test specimen suggested by ISRM is relatively adapted for the assessment of true K_{IIc} [57].

5.1 Processing Methods of AE (Acoustic Emission) Signals

5.1.1 AE Parameter Analysis

The microcrack in solids or structures can develop during the loading processes, causing the accumulated elastic energy to be released quickly, thereby resulting in the AEs (acoustic emissions). This indicates that the AE signal signatures are considerably correlated with the initiation, growth, and intersection of cracking as well as the type and magnitude of microcracks. Generally, the tension crack occurs in a longitudinal wave manner, and the shearing crack happens in a shear wave manner. Further, the longitudinal wave propagation precedes that of shear wave. Hence, the mode-I (tension) crack will generate the AE waveform with shorter rise time and higher frequency, whereas the mode-II (shearing) crack will produce the AE waveform with longer rise time and lower frequency. Clearly, the above-mentioned failure mechanisms can be synthesized by the *RA–AF* distribution, as displayed in Fig. 5.2. The two significant AE indices (*RA* and *AF*) can be estimated via Eqs. (5.1) and (5.2) [21, 58]:

$$RA = RT/A \tag{5.1}$$

$$AF = AC/D \tag{5.2}$$

where *RA* denotes the rising angle (ms/v), *RT* stands for the rising time (μs), *AF* symbolizes the average frequency (kHz), *AC* marks the AE counts, *D* indicates the durative time (μs), and *A* means the maximum amplitude (dB). To standardize their units, Eq. (5.1) should be rewritten as follows [21]:

$$RA = RT/10^{A/20-1} \tag{5.3}$$

Fig. 5.2 Conventional crack identification method based on the *RA* and *AF* values

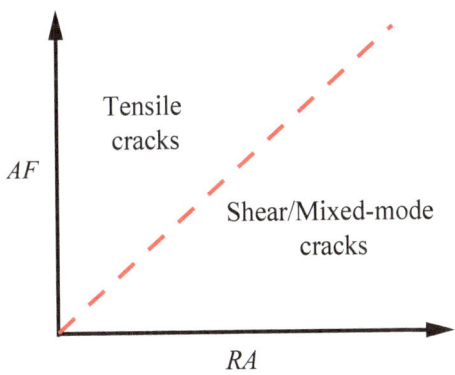

5.1.2   KDE (Kernel Density Estimation) Method

The KDE (Kernel density estimation) method provides a systematically statistical procedure for evaluating the damage and failure degree in structure and machine. Note that this methodology has been gradually emphasized in civil engineering to characterize the density distribution of AE signals. The essential principle of this KDE algorithm is that each AE data contributes a probability density "atom" to the estimation [59], and other introductions are omitted for the sake of brevity.

5.1.3   AE Spectrum Analysis

The AE spectrum analysis can effectively reveal the mechanical mechanisms based on the collected AE signals of deformation processes. Since the FFT (fast Fourier transformation) method is superior in analyzing nonstationary AE signals, each two-dimensional AE spectrum can be extracted from the corresponding time domain waveform of AE signals in conjunction with the MATLAB programming, and the AE dominant frequency is interpreted as the frequency pertaining to the highest amplitude of extracted AE spectrums (see Fig. 5.3). The computed principles of FFT are briefly introduced as below [60]:

$$X(k) = \sum_{n=0}^{N-1} x(n) W_N^{nk} \tag{5.4}$$

$$W_N = e^{-2\pi j/N} \tag{5.5}$$

$$W_N^{k+N/2} = -W_N^k \tag{5.6}$$

$$W_N^{n(N-k)} = W_N^{k(N-n)} = W_N^{-nk} \tag{5.7}$$

where $X(k)$ is the input signal sequence in the frequency domain and is composed of N AE data points, k is taken as $0, 1, \ldots, N-1$, $x(n)$ is the original signal sequence in the time domain, n is the number of signal sequence, and W_N is the twiddle coefficient.

5.1.4   Hierarchical Clustering Algorithm

The HC (Hierarchical Clustering) algorithm is an unsupervised approach to machine learning. The fundamental principle of HC is to evaluate the similarity among tested data points by computing their pairwise distances, followed by constructing a nested

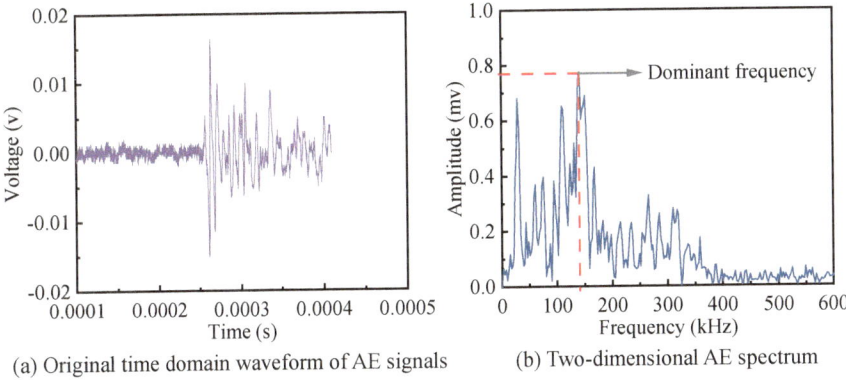

(a) Original time domain waveform of AE signals   (b) Two-dimensional AE spectrum

Fig. 5.3 Extraction of AE spectrum using the FFT (fast Fourier transformation) method

clustering tree with hierarchy according to the degree of similarity. The procedures of HC are generalized as follows [61]:

(1) Each tested data point is taken as a cluster, then the *ED* (Euclidean distance) between pairwise clusters is determined from the following expression:

$$ED = \sqrt{(x_i - x_j)^2 + (y_i - y_j)^2}$$ (5.8)

(2) The two nearest clusters are then merged to create a new cluster.
(3) Continue to perform step (2) iteratively until a solitary cluster encompassing all data points is achieved.

5.2 Experiment Apparatus and Specimen Preparation

The true mode-II fracture experiments are performed on the investigated sandstone specimens via the mechanical testing machine with a compressive capacity of 100 kN, and true mode-II fracture testing procedures are displayed in Fig. 5.4. Meanwhile, the fracture history is captured by the technologically advanced PCI-II AE detector of the Physical Acoustics Corporation, and the monitoring values for AE threshold and preamplifier are respectively adjusted as 30 dB and 40 dB [18], as portrayed in Fig. 5.5. According to the conclusion of Khan and Al-Shayea [62], the variations of fracture resistance were negligible at lower loading rates. Further, the rates of 0.05 ~ 5 mm/min are commonly employed by scientists and investigators for the quasi-static evaluation of fracturing resistance [29]. Hence, this research adopts the invariable rate of 0.3 mm/min to accomplish the true mode-II fracture tests of sandstone specimens. According to the descriptions of previous literature on the geometry and dimension of true mode-II test specimens, the prepared sandstone blocks are processed into specific specimens, as displayed in Table 5.1.

Fig. 5.4 True mode-II fracture testing procedures

Fig. 5.5 PCI-II acoustic
emission detector

Table 5.1 Geometry and dimension of true mode-II test specimens

Specimen type	Diameter (mm)	Height (mm)	Length (mm)	Width (mm)	Thickness (mm)	Notch length (mm)	Ligament length (mm)	Notch direction
SCC	40	80	–	–	–	$R = a = 20$	$C = 16$	Horizontal
SB	–	–	70	70	70	$a = 21$	28	70°
PTS	50	30	–	–	–	$d = 10$	$IP = 10$	Vertical
ZCCDS	50	50	–	–	–	$a = 7.5$	$l_2 = 5$	Vertical

5.3 Experimental Results and Analyses

Figure 5.6 presents the curves of load-deformation for distinct sandstone specimens under pure mode-II loading. The measured curves demonstrate that the rupture of these tested sandstone specimens happens in a brittle manner. Taking the SCC and SB testing methods as examples, the fractured sandstone specimens are displayed in Fig. 5.7. It can be observed from Fig. 5.7 that the investigated sandstone specimens are broken into two identical fragments and the macroscopic rupture surfaces are relatively straight and smooth, indicating that a self-planar cracking propagation pattern appears in each mode-II test specimen. However, the PTS sandstone specimen is broken into two principal parts (internal solid cylinder and hollow centre cylinder). The failure surface on internal solid cylinder between the upper and lower notch

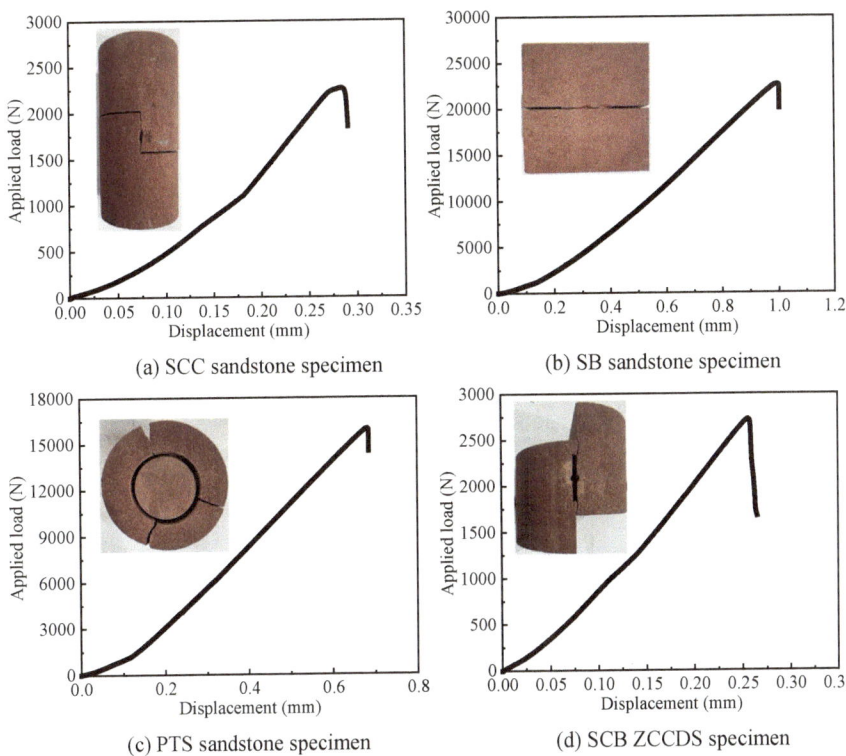

(a) SCC sandstone specimen

(b) SB sandstone specimen

(c) PTS sandstone specimen

(d) SCB ZCCDS specimen

Fig. 5.6 Curves of load-deformation for distinct mode-II test specimens

fronts is comparatively smooth along the vertical direction, implying a self-planar cracking growth by shearing stress takes place in tested PTS bedded shale. While the hollow centre cylinder is composed of fragments, which are separated by almost vertical through-wall fissures (could be formed by the tensile stress). These observations are also reflected in PTS granite specimens tested by Yin et al. [57]. Note that the regression lines presented in Fig. 5.7 are acquired from the box-counting method using the 3D laser scanning technique [], and the slope of the regression lines represents the fractal dimension of macroscopic rupture surfaces. The greater the fractal dimension, the rougher the fracturing surface. Compared with the smallest fractal dimension $D_f = 2$ [63], the D_f values for distinct mode-II test specimens are fairly small, indicating that the rupture of these mode-II test specimens occurs in a self-similar manner.

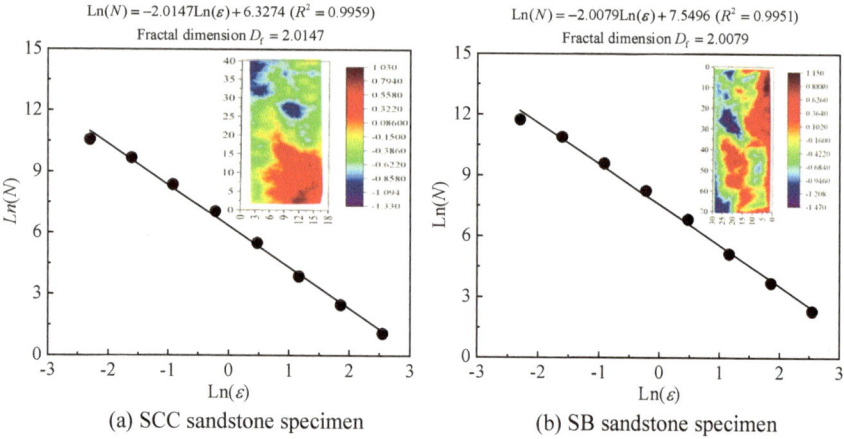

Fig. 5.7 Fractured sandstone specimens under distinct mode-II loadings

5.3.1 Evaluation of Fracture Resistance

To evaluate the fracture resistance of the investigated sandstone under distinct mode-II loadings, this work adopts the two important fracture parameters, namely shear strength and fracture toughness. The true mode-II fracture toughness K_{IIc} can be evaluated by Eqs. (3.5)–(3.9), and the shear strength can be estimated by the following formulas:

$$
\begin{cases}
\tau = \dfrac{P_{\max}}{C \times D} \\[2mm]
\tau = \dfrac{P(\sin\alpha - \tan\psi\cos\alpha)}{BW} \\[2mm]
\tau = \dfrac{P_{\max}}{2\pi \times ID \times IP} \\[2mm]
\tau = \dfrac{P_{\max}}{2l_2 \times D}
\end{cases}
\tag{5.9}
$$

According to Eqs. (3.5)–(3.9) and (5.9), the values of shear strength and true mode-II fracture toughness for the investigated sandstone are presented in Fig. 5.8. One can be concluded from this figure that the average values for τ are 4.11 MPa, 2.21 MPa, 10.31 MPa, and 5.36 MPa relative to the SCC, SB, PTS, and ZCCDS sandstone specimens, respectively, and those for K_{IIc} are 0.51 MPa·m$^{1/2}$, 0.91 MPa·m$^{1/2}$, 1.30 MPa·m$^{1/2}$, and 2.44 MPa·m$^{1/2}$. The true mode-II testing results show that the fracture resistance is dependent on the loading and geometry configurations of test specimens.

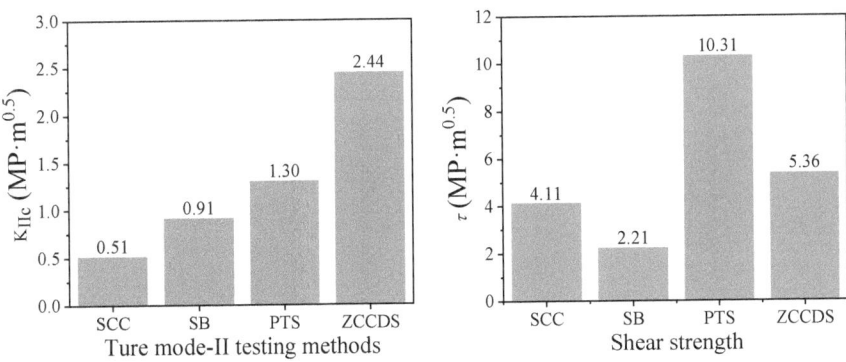

Fig. 5.8 Values of shear strength and true mode-II fracture toughness for the investigated sandstone

5.3.2 Distribution of AE Amplitudes

The AE amplitude can be employed to evaluate the level of energy release in the fracturing processes using a statistical method. The AE amplitudes in the SCC testing are categorized as nine groups: 30–39, 40–44, 45–49, 50–54, 55–59, and 60–99 dB. Then this research adopts the frequency to weight each AE amplitude group, and the distribution of AE amplitudes is formulated by the typical and common Gaussian function in statistics as below [18]:

$$G\left(\overline{dB}\right) = G_0 + A \exp\left[-\left(\overline{dB} - B\right)^2 / 2C^2\right] \qquad (5.10)$$

Where $G\left(\overline{dB}\right)$ denotes the frequency of the corresponding AE amplitude group, $\overline{dB}$ represents the arithmetic mean for each AE amplitude group, and the matching coefficients (A, B, and C) can be acquired from the regression analysis.

Taking the SCC testing method as an example, Fig. 5.9 shows that the Gaussian function can adequately reveal the distribution laws of AE amplitudes for the SCC sandstone. The maximum frequencies appear in 40–49 dB groups for the SCC sandstone. This indicates that these AE signals with aforementioned AE amplitudes are generated by the predominant microcracks within the SCC sandstone and can be interpreted as the characteristic signals of fracturing and damage.

5.3.3 Evolutions of RA and AF Values

Taking the SCC testing method as an example, Fig. 5.10 shows the temporal evolutions of *RA* and *AF* values for the SCC sandstone. Herein, the *RA* and *AF* values are obtained using the moving average of 10 AE events to properly reduce the scatter

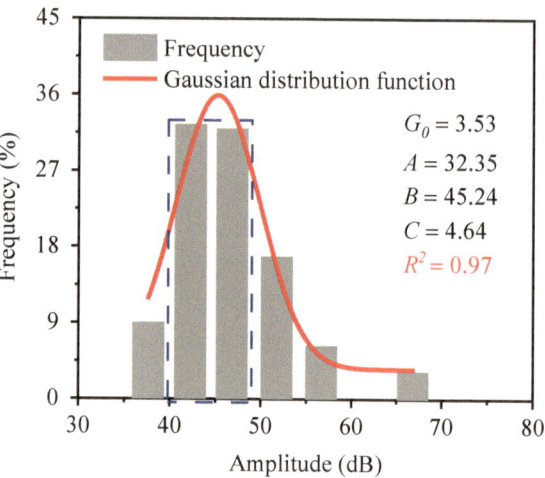

Fig. 5.9 Distributions of AE amplitudes for the SCC sandstone

of AE signal data. It is observed from this figure that there is an opposite development tendency in *RA* and *AF* values, and the phenomenon is consistent with previous research [64]. When relatively higher *RA* values are monitored, indicating that the partial stress drop and large-scale or more intense cracks take place in the tested specimens, which can furnish certain precursory information for engineering failure diagnosis [65]. According to the conventional crack classification criterion (see Fig. 5.1), the tensile and shear/mixed-mode cracks can be recognized using the correlation between *RA* and *AF* (i.e., manually straight line of 45°), as depicted in Fig. 5.11. On can conclude from this figure that the number of tensile cracking is significantly greater than that of shear/mixed-mode cracking. These phenomena also implies that the conventional crack classification criterion can be interpreted as a qualitative analysis rather than a quantitative analysis. One reasonable explanation is that this traditional criterion has not been justified in rocks because it is initially deduced from the exploration of concrete, which is suggested as an approximate and empirical methodology [67–69].

5.4 Discussion

As investigated earlier, the spectrum analysis of AE waveform signals possessed great potential in cracking pattern identification and damage mechanism judgement because the AE waveform can comprehensively reflect the cracking behaviors and energy-releasing level. For various types of rocks (marble, granite, and diorite) under the uniaxial compression, the double characteristic bands of AE dominant frequencies were detected by Zhang et al. [69, 70] as originators who established a quantitatively novel criterion for discerning the failure modes. Specifically, AE signals with H-type spectrums were driven by micro-shear (mode II) failure events, and

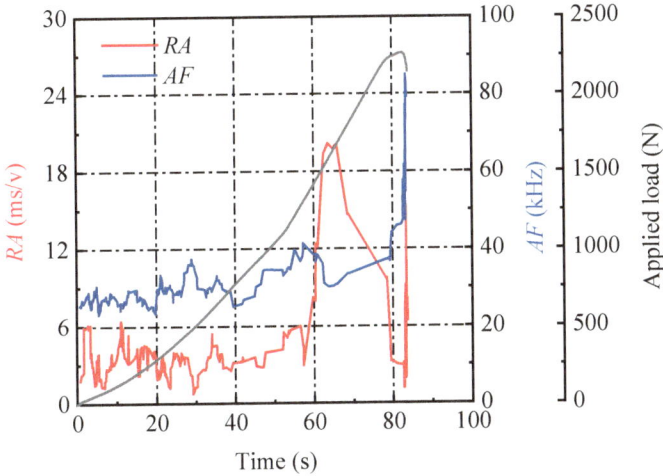

Fig. 5.10 Temporal evolutions of *RA* and *AF* values for the SCC sandstone

Fig. 5.11 Classification results of cracking modes (or failure mechanisms) using the conventional RA-AF method

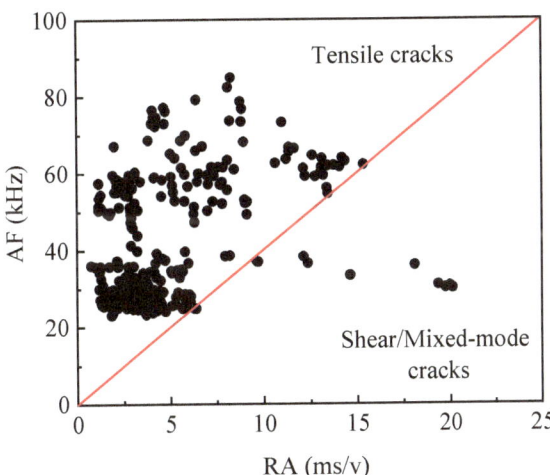

AE signals with L-type spectrums were released by micro-tension (mode I) failure events. Similar phenomena are observed in the PTS testing of the SCC sandstone specimen, as illustrated in Fig. 5.12. According to the distribution signatures of AE dominant frequencies for flawed sandstone under the uniaxial compression, AE dominant frequencies were categorized by Niu and co-workers [60] as three characteristic hierarchies: low (0 ~ 60 kHz), medium (60 ~ 120 kHz), and high (120 ~ 180 kHz). To quantitatively identify the fracturing modes, the AE signals with low frequencies are interpreted as L-type spectrums, and the AE signals with medium and high frequencies are defined as H-type spectrums. For coal specimens under the

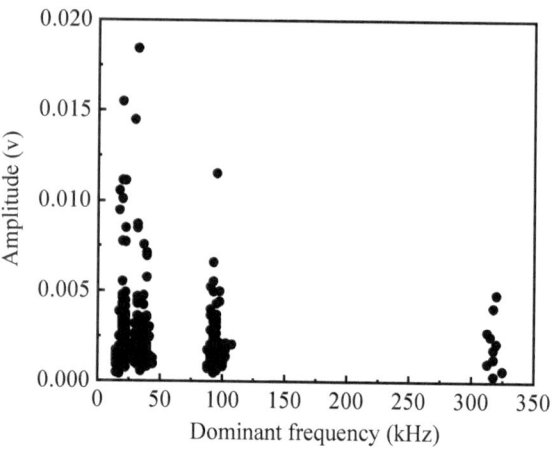

Fig. 5.12 Distributions of AE dominant frequency for the SCC sandstone specimen

uniaxial compression, the peak frequencies of AE signals are identified as four characteristic bands of 0–50 kHz, 50–150 kHz, 150–200 kHz, and 200–300 kHz, which respectively correspond to low (L-type), intermediate, high (H-type), and ultrahigh frequencies [71]. In the direct shear testing, the percentage of AE signals with dominant frequencies around 300 kHz increased considerably and the percentage of AE signals with dominant frequencies below 100 kHz decreased dramatically as the normal load increased [72]. Based on the AE response signatures of the NANO-30 sensor, Guo and co-workers [73, 74] suggested that the frequency corresponding to 1 v was referred to the foundation of partitioning the AE dominant frequency range. To explore the AE signal characteristics of thermally-treated granite under the Brazilian disk tension, the AE dominant frequencies were recognized as three characteristic bands: low (<160 kHz), intermediate (160 ~ 410 kHz), and high (>410 kHz) [73]. In previous literature, the demarcation of AE dominant frequency range was determined in a subjective manner.

The aforementioned phenomena indicate that the magnitude of AE dominant frequency is relative and is significantly influenced by both loading conditions and material types. However, the determination of AE dominant frequency range seems to be questionable because there is no uniform division standard. Considering the hierarchical phenomena of AE dominant frequency distribution, this research develops the HC (hierarchical clustering) algorithm for better AE data classification. According to the correlation between the AE dominant frequency range and damage mechanisms (or cracking modes), the AE dominant frequencies in the SCC testing can be segmented by the developed HC algorithm as three groups: low, intermediate, and high, with corresponding tensile, tensile-shear, and shear cracks, respectively. Consequently, the AE dominant frequency distributions of the SCC sandstone specimen acquired from the HC and KDE methods are portrayed in Fig. 5.13. Although the signals of low and intermediate frequencies are mainly monitored in the whole loading processes of the SCC sandstone specimen, the onset of high dominant frequencies can be generally interpreted as the precursor of macroscopic fracture.

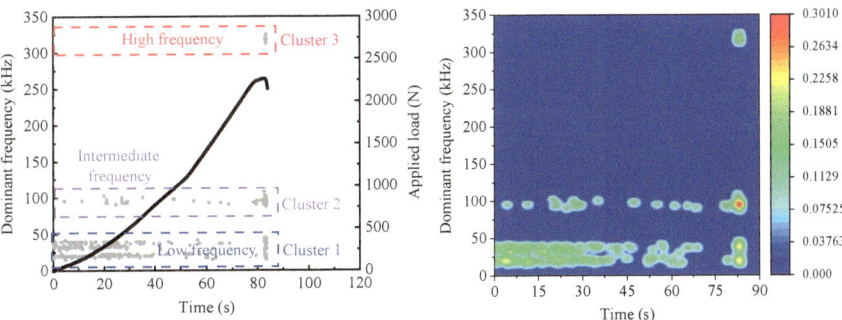

Fig. 5.13 Evaluation of damage mechanisms (or cracking modes) using the frequency-domain-based hierarchical clustering algorithm

In other words, the signals of high dominant frequencies are mostly detected when the macroscopic failure is imminent, and shear cracks dominate the ultimate failure of the SCC sandstone specimen. Accordingly, the proposed HC approach in failure mechanism assessment is superior to the conventional AE index analysis (i.e., the classical *RA-AF* method). What's more, the present frequency-domain-based hierarchical clustering algorithm can independently solve the demarcation problem of AE dominant frequency ranges.

5.5 Conclusion

In this work, the SCC, SB, PTS, and ZCCDS test methods are employed to evaluate the true mode-II fracture properties, and the AE (acoustic emission) monitoring technique is applied to explore the true mode-II fracture mechanisms. The primary conclusions are obtained as follows:

(1) The average values for τ are 4.11 MPa, 2.21 MPa, 10.31 MPa, and 5.36 MPa relative to the SCC, SB, PTS, and ZCCDS sandstone specimens, respectively, and those for K_{IIc} are 0.51 MPa·m$^{1/2}$, 0.91 MPa·m$^{1/2}$, 1.30 MPa·m$^{1/2}$, and 2.44 MPa·m$^{1/2}$. The true mode-II testing results show that the fracture resistance is dependent on the loading and geometry configurations of test specimens.

(2) The Gaussian function can well reveal the distribution characteristics of AE amplitudes, and the maximum frequencies appear in 40–49 dB groups for the SCC sandstone.

(3) Although the signals of low and intermediate frequencies are mainly monitored in the whole loading processes of the SCC sandstone specimen, the onset of high dominant frequencies can be generally interpreted as the precursor of macroscopic fracture.

(4) The dominant frequency-based hierarchical clustering algorithm is adopted in this work to address the demarcation problem of AE dominant frequency ranges, which overcomes the subjective judgment in previous research. The AE dominant frequencies in the SCC testing can be segmented by the developed hierarchical clustering algorithm as three groups: low, intermediate, and high, with corresponding tensile, tensile-shear, and shear cracks, respectively.

References

1. Wang CL, Zhao Y, Ning L, Bi J (2022) Permeability evolution of coal subjected to triaxial compression based on in-situ nuclear magnetic resonance. Int J Rock Mech Min Sci 159:105213
2. Zhao Y, Wang CL, Ning L, Zhao HF, Bi J (2022) Pore and fracture development in coal under stress conditions based on nuclear magnetic resonance and fractal theory. Fuel 309:122112
3. Wang CL, Zhang KP, Zhao Y, Bi J, Ning L, Zhang K (2022) An adsorption model for cylindrical pore and its method to calculate pore size distribution of coal by combining NMR. Chem Eng J 450:138415
4. Zhang YF, Long AF, Zhao Y, Zang A, Wang CL (2023) Mutual impact of true triaxial stress, borehole orientation and bedding inclination on laboratory hydraulic fracturing of Lushan shale. J Rock Mech Geotech Eng
5. Zhao Y, Zhang YF, Yang HQ, Liu Q, Tian GD (2022) Experimental study on relationship between fracture propagation and pumping parameters under constant pressure injection conditions. Fuel 307:121789
6. Zheng K, Wang CL, Zhao Y, Bi J, Liu HF (2023) A criterion of modified three-dimensional mean strain energy density for predicting shale mixed-mode I/III fracture toughness. J Rock Mech Geotech Eng
7. Bidadi J, Aliha MRM, Akbardoost J (2022) Development of maximum tangential strain (MTSN) criterion for prediction of mixed-mode I/III brittle fracture. Int J Solids Struct 256:111979
8. Moghaddam MR, Ayatollahi MR, Berto F (2017) Mixed mode fracture analysis using generalized averaged strain energy density criterion for linear elastic materials. Int J Solids Struct 120:137–145
9. Ayatollahi M, Saboori B (2015) Maximum tangential strain energy density criterion for general mixed mode I/II/III brittle fracture. Int J Damage Mech 24:263–278
10. Ayatollahi MR, Saboori B (2015) T-stress effects in mixed mode I/II/III brittle fracture. Eng Fract Mech 144:32–45
11. Karimi HR, Aliha MRM, Ebneabbasi P, Salehi SM, Khedri E, Haghighatpour PJ (2023) Mode I and mode II fracture toughness and fracture energy of cement concrete containing different percentages of coarse and fine recycled tire rubber granules. Theor Appl Fract Mech 123:103722
12. Khansari NM, Aliha MRM (2023) Mixed-modes (I/III) fracture of aluminum foam based on micromechanics of damage. Int J Damage Mech 32(4):519–548
13. Omidvar N, Aliha MRM, Khoramishad H (2023) Hygrothermal degradation of MWCNT/epoxy brittle materials under I/II combined mode loading conditions: An experimental, micro structural and theoretical study. Theor Appl Fract Mech 125:103896
14. Mehraban MR, Bahrami B, Ayatollahi MR, Nejati M (2023) A non-local XFEM-based methodology for modeling mixed-mode fracturing of anisotropic rocks. Rock Mech Rock Eng 56:895–909
15. Bahrami B, Talebi H, Ayatollahi MR, Khosravani MR (2023) Artificial neural network in prediction of mixed-mode I/II fracture load. Int J Mech Sci 248:108214
16. Sapora A, Ferrian F, Cornetti P, Talebi H, Ayatollahi MR (2023) Ligament size effect in largely cracked tensile structures. Theor Appl Fract Mech 125:103871

17. Bahrami B, Ayatollahi MR, Mehraban MR, Nejati M, Berto F (2022) On the effects of higher order stress terms in pure mode III loading of bi-material notches. Fatigue Fract Eng Mater Struct 45:3333–3346
18. Zheng K, Wang CL, Zhao Y, Bi J, Liu HF (2023) Theoretical and experimental exploration on the combined mode I + III fracture toughness of shale using the edge-notched disk bending method with acoustic emission monitoring. Theor Appl Fract Mech 125:103870
19. Zhao Y, Zheng K, Wang CL, Bi J, Zhang H (2022) Investigation on model-I fracture toughness of sandstone with the structure of typical bedding inclination angles subjected to three-point bending. Theor Appl Fract Mech 119:103327
20. Zheng K, Wang CL, Zhao Y, Bi J (2023) Theoretical and experimental researches on fracture toughness for bedded shale using the centrally cracked Brazilian disk method with acoustic emission monitoring. Theor Appl Fract Mech 124:103784
21. Zheng K, Wang CL, Zhao Y, Bi J, Zhang H (2023) Assessment on anisotropy degree and fracture modes for weakly anisotropic sandstone using the acoustic emission technique. Fatig Fract Eng Mater Struct 1–18
22. Aliha MRM, Hosseinpour GhR, Ayatollahi MR (2013) Application of cracked triangular specimen subjected to three-point bending for investigating fracture behavior of rock materials. Rock Mech Rock Eng 46:1023–1034
23. Jalayer R, Saboori B, Ayatollahi MR (2023) A novel test specimen for mixed mode I/II/III fracture study in brittle materials. Fatig Fract Eng Mater Struct 1–13
24. Bahrami B, Nejati M, Ayatollahi MR, Driesner T (2022) True mode III fracturing of rocks: An axially double-edge notched Brazilian disk test. Rock Mech Rock Eng
25. Aliha MRM, Kosarneshan K, Salehi SM, Haghighatpour PJ, Mousavi A (2023) On the statistical prediction of K_{Ic} and G_{Ic} for railway andesite ballast rock using different three-point bend disc samples. Rock Mech Rock Eng
26. Aliha MRM, Ayatollahi MR, Akbardoost J (2012) Typical upper bound–lower bound mixed mode fracture resistance envelopes for rock material. Rock Mech Rock Eng 45:65–74
27. Pietras D, Aliha MRM, Kucheki HG, Sadowski T (2023) Tensile and tear-type fracture toughness of gypsum material: Direct and indirect testing methods. J Rock Mech Geotech
28. Karimi HR, Bidadi J, Aliha MRM, Mousavi A, Mohammadi MH, Haghighatpour PJ (2023) An experimental study and theoretical evaluation on the effect of specimen geometry and loading configuration on recorded fracture toughness of brittle construction materials. J Build Eng
29. Liu J, Qiao L, Li Y, Li QW, Fan DJ (2022) Experimental study on the quasi-static loading rate dependency of mixed-mode I/II fractures for marble rocks. Theor Appl Fract Mech 121:103431
30. Hou C, Jin XC, Fan XL, Xu R, Wang ZY (2019) A generalized maximum energy release rate criterion for mixed mode fracture analysis of brittle and quasi-brittle materials. Theor Appl Fract Mech 100:78–85
31. Wang H, Li Y, Cao SG, Fantuzzi N, Pan RK, Tian MY, Liu YB, Yang HY (2020) Fracture toughness analysis of HCCD specimens of Longmaxi shale subjected to mixed mode I-II loading. Eng Fract Mec 239:107299
32. Shen Z, Yu HY, Guo LC, Hao LL, Zhu S, Huang K (2023) A modified 3D G-criterion for the prediction of crack propagation under mixed mode I-III loadings. Eng Fract Mech 281:109082
33. Hua W, Li JX, Zhu ZY, Li AQ, Huang JZ, Gan ZQ, Dong SM (2023) A review of mixed mode I-II fracture criteria and their applications in brittle or quasi-brittle fracture analysis. Theor Appl Fract Mech 124:103741
34. Hua W, Li JX, Zhu ZY, Li AQ, Huang JZ, Dong SM (2023) Experimental study on mode I and mode II fracture properties of heated sandstone after two different cooling treatments. Geomech Energy Environ 100448
35. Huang JZ, Hua W, Li DS, Chen X, You XT, Dong SM, Li JX (2023) Effect of confining pressure on the compression-shear fracture properties of sandstone. Theor Appl Fract Mech 124:103763
36. Hua W, Li JX, Gan ZQ, Huang JZ, Dong SM (2022) Degradation response of mode I and mode III fracture resistance of sandstone under wetting–drying cycles with an acidic solution. Theor Appl Fract Mec 122:103661

37. Zhao YX, Sun Z, Gao YR, Wang XL, Song HH (2022) Influence of bedding planes on fracture characteristics of coal under mode II loading. Theor Appl Fract Mec 117:103131
38. Wang W, Teng T (2022) Experimental study on anisotropic fracture characteristics of coal using notched semi-circular bend specimen. Theor Appl Fract Mec 122:103559
39. Shi XS, Zhao YX, Gong S, Wang W, Yao W (2022) Co-effects of bedding planes and loading condition on mode-I fracture toughness of anisotropic rocks. Theor Appl Fract Mec 117:103158
40. Nejati M, Bahrami B, Ayatollahi MR, Driesner T (2021) On the anisotropy of shear fracture toughness in rocks. Theor Appl Fract Mec 113:102946
41. Aminzadeh A, Bahrami B, Ayatollahi MR, Nejati M (2022) On the role of fracture process zone size in specifying fracturing mechanism under dominant mode II loading. Theor Appl Fract Mec 117:103150
42. Bahrami B, Nejati M, Ayatollahi MR (2020) Theory and experiment on true mode II fracturing of rocks. Eng Fract Mech 240:107314
43. Sakha M, Nejati M, Aminzadeh A, Ghouli S, Saar MO, Driesner T (2022) On the validation of mixed-mode I/II crack growth theories for anisotropic rocks. Int J Solids Struct 241:111484
44. Bahrami B, Ghouli S, Nejati M, Ayatollahi MR, Driesner T (2022) Size effect in true mode II fracturing of rocks: Theory and experiment. Eur J Mech A Solid 94:104593
45. Fan ZD, Xie HP, Zhang R, Lu HJ, Zhou Q, Nie XF, Luo Y, Ren L (2022) Characterization of anisotropic mode II fracture behaviors of a typical layered rock combining AE and DIC techniques. Eng Fract Mech 271:108599
46. Fan ZD, Ren L, Xie HP, Zhang R, Li CB, Lu HJ, Zhang AL, Zhou Q, Ling WQ (2023) 3D anisotropy in shear failure of a typical shale. Petroleum Sci 20:212−229
47. Fan ZD, Xie HP, Ren L, Zhang R, He R, Li CB, Zhang ZT, Wang J, Xie J (2022) Anisotropy in shear-sliding fracture behavior of layered shale under different normal stress conditions. J Cent South Univ 29(11):3678–3694
48. Ma Y, Rao QH, Huang DY, Liu ZL, Yi W, Li P (2023) Gas-mechanical coupled crack initiation analysis for local air-leakage of compressed air energy storage (CAES) cavern with consideration of seepage effect. Theor Appl Fract Mech 125:103827
49. Sun DL, Rao QH, Wang SY, Shen QQ, Yi W (2021) Shear fracture (Mode II) toughness measurement of anisotropic rock. Theor Appl Fract Mec 115:103043
50. Alneasan M, Alzo'ubi AK, Okasha N (2023) A comprehensive study for the effect of sample geometry and lateral pressure on shear fractures using the short core in compression (SCC) method. Eur J Mech A Solid 100:104988
51. Zhang CX, Li DY, Ma JY, Zhu QQ, Luo PK, Chen YD, Han MG (2023) Dynamic shear fracture behavior of rocks: insights from three-dimensional digital image correlation technique. Eng Fract Mech 277:109010
52. Cao RH, Fang L, Qiu XY, Lin H, Li XL, Li WX, Qiao QQ (2023) Effect of heating–water cooling cycle treatment on the pore structure and shear fracture characteristics of granite. Eng Fract Mech 109263
53. Yao W, Wang JX, Wu BB, Xu Y, Xia KW (2023) Dynamic mode II fracture toughness of rocks subjected to various in situ stress conditions. Rock Mech Rock Eng 56:2293–2310
54. Yao W, Xu Y, Xia KW, Wang S (2020) Dynamic mode II fracture toughness of rocks subjected to confining pressure. Rock Mech Rock Eng 53:569–586
55. Backers T, Stephansson O (2012) ISRM Suggested method for the determination of mode ii fracture toughness. Rock Mech Rock Eng 45(6):1011–1022
56. Rao QH, Sun ZQ, Stephansson O, Li CL, Stillborg B (2003) Shear fracture (Mode II) of brittle rock. Int J Rock Mech Min Sci 40:355–375
57. Yin TB, Tan XS, Wu Y, Yang Z, Li MJ (2021) Temperature dependences and rate effects on Mode II fracture toughness determined by punch-through shear technique for granite. Theor Appl Fract Mech 114:103029
58. Wang Y, Zhang B, Gao SH, Li CH (2021) Investigation on the effect of freeze-thaw on fracture mode classification in marble subjected to multi-level cyclic loads. Theor Appl Fract Mech 111:102847

59. Rippengill S, Worden K, Holford KM, Pullin R (2003) Automatic classification of acoustic emission patterns. Strain 39(1):31–41
60. Niu Y, Zhou XP, Berto F (2020) Temporal dominant frequency evolution characteristics during the fracture process of flawed red sandstone. Theor Appl Fract Mech 110:102838
61. Jia JQ, Ju SY (2023) Clustering-based method for locating critical slip surface using the strength reduction method. Comput Geotech 155:105241
62. Khan K, Al-Shayea NA (2000) Effect of specimen geometry and testing method on mixed mode I-II fracture toughness of a limestone rock from Saudi Arabia. Rock Mech Rock Eng 33(3):179–206
63. Ai T, Zhang R, Zhou HW, Pei JL (2014) Box-counting methods to directly estimate the fractal dimension of a rock surface. Appl Surface Sci 314:610–621
64. Aggelis DG (2011) Classification of cracking mode in concrete by acoustic emission parameters. Mech Res Commun 38(3):153–157
65. Zhou XP, Peng SL, Zhang JZ, Zhou JN, Berto F (2022) Experimental study on cracking behaviors of coarse and fine sandstone containing two flaws under biaxial compression. Fatigue Fract Eng Mater Struct 45:2595–2612
66. Ge ZL, Sun Q (2021) Acoustic emission characteristics of gabbro after microwave heating. Int J Rock Mech Min Sci 138:104616
67. Li PF, Sun Q, Geng JS, Jing XD, Tang LY (2023) Study on the characteristics of radon exhalation from rocks in coal fire area based on the evolution of pore structure. Sci. Total Environment 862:160865
68. Farhidzadeh A, Dehghan-Niri E, Salamone S, Luna B, Whittaker A (2013) Monitoring crack propagation in reinforced concrete shear walls by acoustic emission. J Struct Eng 139(12):04013010
69. Zhang ZH, Deng JH (2020) A new method for determining the crack classification criterion in acoustic emission parameter analysis. Int J Rock Mech Min Sci 130:104323
70. Zhang ZH, Ma K, Li H, He ZL (2022) Microscopic investigation of rock direct tensile failure based on statistical analysis of acoustic emission waveforms. Rock Mech Rock Eng 55:2445–2458
71. Dai JJ, Liu JF, Zhou LL, He X (2023) Crack pattern recognition based on acoustic emission waveform features. Rock Mech Rock Eng 56:1063–1076
72. Du K, Li XF, Wang SY, Tao M, Li G, Wang SF (2021) Compression-shear failure properties and acoustic emission (AE) characteristics of rocks in variable angle shear and direct shear tests. Measurement 183:109814
73. Guo P, Wu SC, Zhang G, Chu CQ (2021) Effects of thermally-induced cracks on acoustic emission characteristics of granite under tensile conditions. Int J Rock Mech Min Sci 144:104820
74. Zhao YS, Chen CC, Wu SC, Guo P, Li BL (2022) Effects of 2D & 3D nonparallel flaws on failure characteristics of brittle rock-like samples under uniaxial compression: Insights from acoustic emission and DIC monitoring. Theor Appl Fract Mech 120:103391

Chapter 6
Mixed-Mode I/III Fracture

Shale reservoirs are characterized by both ultralow porosity and permeability, the attached shale gas can be effectively extracted by hydraulic fracturing (also known as fracking) technique to optimize the energy structure [1–8]. The fracking technique holds great potential to enhance the productivity and recovery of shale gas by creating high-conductivity fissures, and a thorough comprehension of fracturing network formation is imperative [9–14]. The understanding of fracture mechanics properties for rock masses is certainly required, especially for the determination of fracture toughness as an important engineering parameter. For actual rock masses with randomly internal cracks, there are three essential fracturing types, namely pure mode I (tension), pure mode II (shear), and pure mode III (tearing) which generate the opening, planar sliding, and nonplanar sliding deformations of cracks respectively [15, 16]. Irwin [17] presented the definition of SIF (stress intensity factor) in 1957. When the SIF attained its critical level, the cracking became unstable and propagated with a rapid velocity, and the critical SIF could powerfully affect the geometries of hydraulic fractures [18]. As a significant fracturing indicator, the fracture toughness of rocks weights the stress and displacement fields at the adjacency of crack fronts and represents the capacity to resist cracking propagation. To estimate the fracture resistances of pure mode I (tension), combined-mode I/II, and pure mode II (shear), distinct loading configurations have been devised and developed. Modeling these fracture types is comparatively straightforward via corresponding test specimens, including centrally cracked Brazilian disk (CCBD), centrally cracked ring disk under diametral compression (CCRDDC), semicircle bending (SCB), edge-cracked triangle under bending (ECTB), and edge-cracked rectangle beam under bending (ECRBB) [19–26]. Additionally, a hexapod testing configuration was specially designed to explore the cracking behaviors of combined modes (I/II and I/III) [27, 28]. In fact, the signatures of hydraulic and natural fractures are multidimensional and multiscale, and the resultant three-dimensional fracturing problems deserve to be emphasized [29]. For this situation, estimating the combined-mode I/III fracture resistance is helpful in analyzing and optimizing the fracture

Y. Zhao et al., *Rock Fracture Mechanics and Fracture Criteria*,
https://doi.org/10.1007/978-981-97-5822-7_6

networks of shale gas reservoirs. Compared to ductile metal materials [30, 31], the investigations of pure mode III and combined-mode I/III for brittle geo-materials are more limited due to the challenges in applying torsional or non-coplanar forces required for introducing mode III, as well as the issues with specimen fabrication [32–34].

Compared with the combined-mode I/II testing, fewer qualified loading configurations are available for the combined-mode I/III testing of geo-materials.

For instance, the pure mode I and mixed-mode I/III fracture behaviors of epoxy resin were explored by Ahmadi-Moghadam and Taheri [35] using the INRBB (inclined-notched rectangular beam bending) testing method that is incapable of providing the pure mode III loading. Similarly, the INSCB (inclined-notched semicircular bending) testing configuration with symmetrical bottom supports was suggested by Pirmohammad and Kiani [36] to assess the pure mode I and mixed mode I + III fracture resistance of HMA concrete. Since this INSCB testing fixture is incompetent to achieve the dominant mode III and pure mode III loading, the other INSCB testing configuration with asymmetrical bottom supports was developed by Bakhshizadeh and Pirmohammad [37] to estimate the pure mode III and mixed mode I + III fracture resistance of marble. However, the latter INSCB loading fixture cannot provide the pure mode I loading. The edge-notched disk bending (ENDB) specimen was established by Tutluoglu and Keles [38] as the forerunners who investigated the pure mode-I fracture resistance of andesite. Subsequently, Aliha et al. [39] successfully extended the ENDB testing technique to permit the gamut of brittle fractures from pure tension to pure torsion or tearing. Under the inspiration of the ENDB specimen, Aliha and co-workers [40] proposed the ENDC (edge-notched disk compression) specimen to compare and measure the mixed-mode I/III fracture resistance of granite. Although the geometry of the edge-notched disk under diametral compression (ENDDC) specimen is identical with that of the ENDB specimen, the mode mixities are considerably susceptible to the orientations of the pre-existing notch, rendering the combined-mode I/III testing more challenging [40]. Further, both in-situ stresses and natural fractures have remarkable influences on the geometry and direction of hydraulic fractures. In tectonic stress regimes, hydraulic fractures initiated from the pre-existing wellbore when the principal stress surpassed the rock's tension strength, then propagated for a shorter distance. Subsequently, these induced fissures manifested as a twisted out-of-plane feature because of the discrepancy among three principal stresses [41]. Undoubtedly, the combined-mode I/III fracture resistance can be estimated via the qualified ENDB specimen with simple geometry, and the corresponding loading apparatus is a common and straightforward three-point bending configuration. In addition, the nature of the fractured ENDB specimen is also consistent with that of hydraulic fractures [42].

Currently, the investigators and practitioners emphasize the pure mode-I fracture resistance measurement of rocks. In particular, the jointed shale reservoirs generally suffer from complicated combined-mode deformations, and the resultant three-dimensional fracturing problems need to be emphasized. As documented earlier, there is a comparatively lacking investigation on the combined-mode fracturing behaviors of shale, especially the determination of pure mode-III and combined-mode I/III fracture toughness. To analyze the fracture mechanisms, this work adopts the published failure criteria: the stress-based three-dimensional maximum tangential stress (3D-MTS) criterion, the strain-based three-dimensional maximum tangential strain (3D-MTSN) criterion, and the energy-based three-dimensional maximum tangential strain energy density factor (3D-MTSEDF) criterion. Then this work presents the modified three-dimensional mean strain energy density criterion. The ENDB sandstone specimens and DENDC sandstone specimens are adopted to implement the pure mode-I, combined-mode I/III, and pure mode-III tests. Further, these established fracture models are applied as contrastive analyses. Finally, the proposed fracture criterion is discussed and validated by the laboratory data.

6.1 Development of the Modified Three-Dimensional Mean Strain Energy Density Criterion

For the combined-mode I/III crack problems, one can conclude from the description of Ayatollahi and co-workers [43, 44] on strain energy density that

$$\frac{dW}{dV} = \frac{1}{2E}\left(\sigma_{r'r'}^2 + \sigma_{\theta'\theta'}^2 + \sigma_{z'z'}^2\right)$$

$$-\frac{\nu}{E}\left(\sigma_{r'r'}\sigma_{\theta'\theta'} + \sigma_{r'r'}\sigma_{z'z'} + \sigma_{z'z'}\sigma_{\theta'\theta'}\right) + \frac{1}{2G}\left(\sigma_{r'\theta'}^2 + \sigma_{r'\theta'}^2 + \sigma_{\theta'z'}^2\right) \quad (6.1)$$

where dW/dV symbolizes the strain energy density in an element, E and $G = E/2(1+\nu)$ mean respectively the Young and shear moduli (ν represents the Poisson coefficient). $\sigma_{r'r'}$, $\sigma_{\theta'\theta'}$, $\sigma_{z'z'}$, $\sigma_{r'\theta'}$, $\sigma_{r'z'}$, and $\sigma_{\theta'z'}$ denote the stress components in the system of the transmitted (r', θ', z') coordinate (see Fig. 2.4) and can be formulated by employing the transformation between the (r, θ, z) and (r', θ', z') coordinates as

$$\begin{cases} \sigma_{r'r'} = \sigma_{rr} \\ \sigma_{\theta'\theta'} = \sigma_{\theta\theta}\cos^2\varphi - \sigma_{\theta z}\sin 2\varphi + \sigma_{zz}\sin^2\varphi \\ \sigma_{z'z'} = \sigma_{\theta\theta}\sin^2\varphi + \sigma_{\theta z}\sin 2\varphi + \sigma_{zz}\cos^2\varphi \\ \sigma_{r'\theta'} = \sigma_{r\theta}\cos\varphi - \sigma_{rz}\sin\varphi \\ \sigma_{r'z'} = \sigma_{r\theta}\sin\varphi + \sigma_{rz}\cos\varphi \\ \sigma_{\theta'z'} = \frac{1}{2}\sigma_{\theta\theta}\sin 2\varphi + \sigma_{\theta z}\cos 2\varphi - \frac{1}{2}\sigma_{zz}\sin 2\varphi \end{cases} \quad (6.2)$$

where the stress components $(\sigma_{rr}, \sigma_{\theta\theta}, \sigma_{zz}, \sigma_{rz}, \sigma_{r\theta}, \text{ and } \sigma_{\theta z})$ in the system of the (r, θ, z) coordinate (see Fig. 2.4) are stipulated as [45]

$$\begin{cases} \sigma_{rr} = \dfrac{K_I}{\sqrt{2\pi r}}\left[\dfrac{5}{4}\cos\dfrac{\theta}{2} - \dfrac{1}{4}\cos\dfrac{3\theta}{2}\right] \\[2mm] \sigma_{\theta\theta} = \dfrac{K_I}{\sqrt{2\pi r}}\left[\dfrac{3}{4}\cos\dfrac{\theta}{2} + \dfrac{1}{4}\cos\dfrac{3\theta}{2}\right] \\[2mm] \sigma_{r\theta} = \dfrac{K_I}{\sqrt{2\pi r}}\left[\dfrac{1}{4}\sin\dfrac{\theta}{2} + \dfrac{1}{4}\sin\dfrac{3\theta}{2}\right] \\[2mm] \sigma_{rz} = \dfrac{K_{III}}{\sqrt{2\pi r}}\sin\dfrac{\theta}{2} \\[2mm] \sigma_{\theta z} = \dfrac{K_{III}}{\sqrt{2\pi r}}\cos\dfrac{\theta}{2} \\[2mm] \sigma_{zz} = v(\sigma_{rr} + \sigma_{\theta\theta}) \end{cases} \tag{6.3}$$

The strain energy within the plastic or damaged zone of radius r_c surrounding the fracture front can be evaluated via the integration of SED relative to r_c as below:

$$E(r_c) = \int_A \frac{dW}{dV}dA = \int_0^{r_c}\int_{-\pi}^{\pi}\frac{dW}{dV}rdrd\theta \tag{6.4}$$

Accordingly, the mean strain energy on the characteristic zone area is assessed as

$$\overline{E(r_c)} = \frac{E(r_c)}{\pi r_c^2} \tag{6.5}$$

For the mixed-mode I/III fracture problems (the planar fracture deflection angle $\theta_c = 0$ [45]), the theoretical fracture toughness ratios extracted from the aforementioned formulae in this paper can be developed to forecast the fracturing initiation in the normalized form:

$$\begin{cases} \dfrac{K_I}{K_{Ic}} = \left(\sqrt{1 + \dfrac{1+v}{1-\kappa}\dfrac{K_{III}^2}{K_I^2}}\right)^{-1} \\[4mm] \dfrac{K_{III}}{K_{Ic}} = \left(\sqrt{\dfrac{K_I^2}{K_{III}^2} + \dfrac{1+v}{1-\kappa}}\right)^{-1} \end{cases} \tag{6.6}$$

in which κ marks an elastic parameter as the function of v and is taken as v for the plane stress state and $(v + 2v^2)$ for the plane strain state.

One can recognize that $\overline{E(r_c)}$ can be decomposed into two significant components, namely volumetric and distortional MSEDs, $\overline{E_v(r_c)}$ and $\overline{E_d(r_c)}$, are respectively defined as follows [46]:

$$\overline{E_v(r_c)} = \frac{10}{50Er_c}(1 - 2v)(1 + v)^2 K_I^2 \tag{6.7}$$

$$\overline{E_d(r_c)} = \frac{10}{50Er_c}\left[(1-2v)(1+v)\left(\frac{6}{10}-v\right)K_I^2 + \frac{16}{10}(1+v)K_{III}^2\right] \tag{6.8}$$

Under the inspiration of F-criterion (see Eq. (6.9) [47]), we consider different weights given to $\overline{E_v(r_c)}$ and $\overline{E_d(r_c)}$ to make an improvement for the conventional 2D-MSED criterion [48] by distinguishing the influences of volumetric and distortional MSEDs as Eq. (6.10).

$$F = \frac{G_I}{G_{Ic}} + \frac{G_{II}}{G_{IIc}} \tag{6.9}$$

$$Z = \frac{\overline{E_v(r_c)}}{\overline{E_{vc}}} + \frac{\overline{E_d(r_c)}}{\overline{E_{dc}}} \tag{6.10}$$

where G_I and G_{II} are respectively the energy release rates (ERRs) of modes I and II, the subscript c symbolizes the critical status, and Z is the new fracture factor introduced in this article.

To characterize the influence of this discrepancy between $\overline{E_{dc}}$ and $\overline{E_{vc}}$ on the improved MSED criterion, a pivotal parameter λ, i.e., the ratio of critical volumetric to distortional MSEDs, is proposed as

$$\lambda = \frac{\overline{E_{vc}}}{\overline{E_{dc}}} \tag{6.11}$$

Then Eq. (6.10) can be converted into

$$\begin{aligned} Z &= \frac{1}{\overline{E_{vc}}}\left[\overline{(E_v(r_c))} + \lambda\overline{E_d(r_c)}\right] \\ &= \frac{1}{\overline{E_{vc}}}\left\{\frac{10}{50Er_c}(1-2v)(1+v)^2K_I^2 + \lambda\frac{10}{50Er_c}\left[(1-2v)(1+v)\left(\frac{6}{10}-v\right)K_I^2 + \frac{16}{10}(1+v)K_{III}^2\right]\right\} \end{aligned} \tag{6.12}$$

Considering a special mode-I fracturing case with $K_{III} = 0$ and $K_I = K_{Ic}$, then Eq. (6.12) can be simplied as

$$Z = \frac{1}{\overline{E_{vc}}}\left[\frac{10}{50Er_c}(1-2v)(1+v)^2K_{Ic}^2 + \lambda\frac{10}{50Er_c}(1+v)(1-2v)\left(\frac{6}{10}-v\right)K_{Ic}^2\right] \tag{6.13}$$

Under the specific mode-III fracturing case with $K_{III} = K_{IIIc}$ and $K_I = 0$, combining Eqs. (6.12) and (6.13) yields the ratio of K_{IIIc} to K_{Ic}:

$$\frac{K_{IIIc}}{K_{Ic}} = \sqrt{\frac{(1-2v)(1+v)^2 + \lambda(1-2v)(1+v)(0.6-v)}{1.6\lambda(1+v)}} \tag{6.14}$$

For the general combined-mode I/III fracturing case, introducing Eq. (6.13) into Eq. (6.12) yields

$$\left[(1-2v)(1+v)^2 K_{Ic}^2 + \lambda(1+v)(1-2v)(0.6-v)K_{Ic}^2\right]$$
$$= \left\{(1-2v)(1+v)^2 K_I^2 + \lambda[(1+v)(1-2v)(0.6-v)K_I^2 + 1.6(1+v)K_{III}^2]\right\}$$
$$(6.15)$$

When both sides of Eq. (6.15) are divided by K_I^2, K_{III}^2 respectively, the normalized fracture toughness, K_I/K_{Ic} and K_{III}/K_{Ic}, can be extracted to forecast the onset of combined-mode I/III fracturing as follows:

$$\begin{cases} \dfrac{K_{Ic}^2}{K_I^2} = \dfrac{(1-2v)(1+v)^2 + \lambda\left[(1+v)(1-2v)(0.6-v) + 1.6(1+v)\frac{K_{III}^2}{K_I^2}\right]}{(1-2v)(1+v)^2 + \lambda(1+v)(1-2v)(0.6-v)} \\[4mm] \dfrac{K_{Ic}^2}{K_{III}^2} = \dfrac{(1-2v)(1+v)^2\frac{K_I^2}{K_{III}^2} + \lambda\left[(1+v)(1-2v)(0.6-v)\frac{K_I^2}{K_{III}^2} + 1.6(1+v)\right]}{(1-2v)(1+v)^2 + \lambda(1+v)(1-2v)(0.6-v)} \end{cases}$$
$$(6.16)$$

Note that Eqs. (6.14) and (6.16) represent the mathematical forms of the modified three-dimensional MSED criterion under plane strain status. Similarly, the corresponding expressions for plane stress status can be derived as Eqs. (6.17) and (6.18):

$$\frac{K_{IIIc}}{K_{Ic}} = \sqrt{\frac{5(1-2v) + \lambda(3+2v)}{8\lambda(1+v)}} \qquad (6.17)$$

$$\begin{cases} \dfrac{K_{Ic}^2}{K_I^2} = \dfrac{5(1-2v) + \lambda\left[(3+2v) + 8(v+1)\frac{K_{III}^2}{K_I^2}\right]}{5(1-2v) + \lambda(3+2v)} \\[4mm] \dfrac{K_{Ic}^2}{K_{III}^2} = \dfrac{5(1-2v)\frac{K_I^2}{K_{III}^2} + \lambda\left[(3+2v)\frac{K_I^2}{K_{III}^2} + 8(v+1)\right]}{5(1-2v) + \lambda(3+2v)} \end{cases} \qquad (6.18)$$

The following steps are utilized for forecasting the onset of combined-mode I/III fracturing:

(1) The experimental fracture resistance (K_I and K_{III}) values are acquired from related expressions (e.g., Eqs. (3.10)–(3.14)), then the calculated K_I and K_{III} values are all divided by K_{Ic} (pure mode-I fracture resistance) to obtain the experimental fracture resistance ratios (K_I/K_{Ic} and K_{III}/K_{Ic}).

(2) The non-coplanar fracturing twist angle φ_c is solved by Eq. (2.67), then substituting the calculated φ_c values into Eqs. (2.70) (3D-MTS crietrion), (2.71) (3D-MTSN criterion), and (2.72) (3D-MTSEDF criterion) yields the corresponding fracture resistance envelopes.

(3) The critical volumetric to distortional MSED ratio λ can be determined by introducing the tested K_{Ic} and K_{IIIc} values into Eq. (6.17), then the present fracture resistance envelopes can be provided by substituting the obtained λ value into Eq. (6.18).
(4) Consequently, both experimental data points and theoretical fracture resistance envelopes are outlined in a $(K_I/K_{Ic}) - (K_{III}/K_{Ic})$ diagram.

6.2 Experiment Apparatus and Specimen Preparation

The combined-mode I/III loading experiments are implemented on the investigated sandstone specimens using a mechanical tester of DANA company with a compressive capacity of 100 kN, and the loading procedures are displayed in Fig. 6.1. One could conclude from previous literature that lower loading rates had insignificant influences on the fracture resistance measurements for rock materials. In addition, a relatively constant fracture resistance is expected to be determined because it is recognized as an independently inherent material property. In the current study, the ENDB sandstone specimens and DENDC sandstone specimens are respectively loaded at a constant displacement rate of 0.1 and 0.3 mm/min to appropriately eliminate the loading rate effect on the combined-mode I/III fracture resistance.

The investigated sandstone blocks come from Zigong, Sichuan Province. Based on the investigation of Aliha and co-workers [49], prepared sandstone blocks are processed into the unnotched Brazilian disk specimens with a radius of $R = 37.5$ mm and a height of $t = 40$ mm. Then an edge-straight notch with the depth of $a = 24$ mm and the interval of $d = 1$ mm is processed in the center of unnotched samples along the diametrical orientation. According to the loading angle β shown in Fig. 3.9,

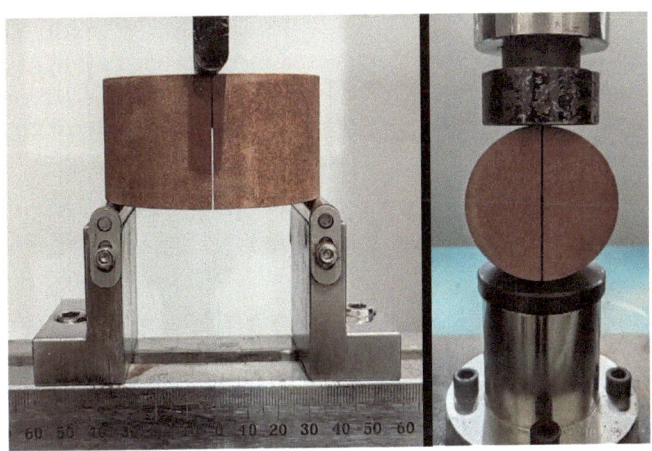

(a) ENDB sandstone specimen   (b) DENDC sandstone specimen

Fig. 6.1 Loading procedures

Table 6.1 Dimensionless fracture parameters of the ENDB specimen for $a/B = 0.6$ and $S/R = 0.925$

Loading mode	Loading angle β ($°$)	Mode mixity parameter M^e	Mode-I geometry factor F_I	Mode-III geometry factor F_{III}
Pure mode I	0	1	0.477	0
Mixed-mode I/III	50	0.66	0.093	0.056
Pure mode III	62.5	0	0	0.050

Table 6.2 Dimensionless fracture parameters of the DENDC specimen for $a/B = 0.6$

Loading mode	Loading angle β ($°$)	Mode mixity parameter M^e	Mode-I geometry factor F_I	Mode-III geometry factor F_{III}
Pure mode I	0	1	0.39	0
Mixed-mode I/III	9	0.5	0.32	0.32
Pure mode III	27	0	0	0.63

pure mode I (i.e., $\beta = 0°$), combined-mode I/III (i.e., $\beta = 50°$), and pure mode III (i.e., $\beta = 62.5°$) can be achieved. Based on the investigated results of Aliha and co-workers [49], the specific dimensionless fracture parameters (F_I and F_{III}) of the ENDB specimen for $a/B = 0.6$ and $S/R = 0.925$ are provided in Table 6.1.

According to the investigation of Aliha and co-workers [50], prepared sandstone blocks are machined into the unnotched Brazilian disk specimens with a radius of $R = 37.5$ mm and a height of $t = 40$ mm. Then two edge-straight notches with the depth of $a/2 = 12$ mm and the interval of $d = 1$ mm is processed in the center of unnotched samples along the diametrical orientation. According to the loading angle β shown in Fig. 3.11, pure mode I (i.e., $\beta = 0°$), combined-mode I/III (i.e., $9°$), and pure mode III (i.e., $\beta = 27°$) can be achieved. Based on the investigated results of Aliha and co-workers [50], the specific dimensionless fracture parameters (F_I and F_{III}) of the DENDC specimen for $a/B = 0.6$ are provided in Table 6.2.

6.3 Experiment Results and Analyses

6.3.1 Analyses of Peak Load and Applied Energy

Figures 6.2 and 6.3 reports the evolutions of applied load versus loading point deformation for the investigated ENDB and DENDC sandstone specimens under the specific loading mode mixities. Demonstrably, the measured curves for each ENDB shale specimen are characterized by three classical phases: (1) negligible nonlinearity in the initial deformation, (2) linear elasticity with the incremental force, and

(3) ultimately sudden instability with a brittle cracking. One can determine from Figs. 6.6.4, 5 that the magnitude of peak load p_{max} for the investigated ENDB and DENDC sandstone specimens is remarkably enhanced by the transition from pure mode-I loading to pure mode-III loading. When the loading mode mixity index M^e = 1, 0.66, and 0, the average p_{max} values for the ENDB sandstone specimens are respectively 744.2, 1906.4, and 3436.4 N. When the loading mode mixity index M^e = 1, 0.5, and 0, the average p_{max} values for the DENDC sandstone specimens are respectively 5150.6, 5159.9, and 8858.7 N. By conducting combined-mode I/III tests on the ENDB and DENDC sandstone specimens, the resulting work can be interpreted as the applied energy or input energy E_f, which is determined by computing the area below the measured curve (up to the onset of peak load P_{max}), as portrayed in Fig. 6.4a [51–53]. Particularly, the the applied energy or input energy E_f denotes the energy required for fragmenting rocks, which can be considered as the candidate fracture parameter for rock engineering applications. Figure 6.4b illustrates the evolution of E_f with the increasing loading mode mixity index M^e. When the loading mode mixity index M^e = 1 (i.e., pure mode-I), 0.66, and 0 (i.e., pure mode-III), the average E_f values for the ENDB sandstone specimens are respectively 0.07, 0.21, and 0.31 J. When the loading mode mixity index M^e = 1 (i.e., pure mode-I), 0.5, and 0 (i.e., pure mode-III), the average E_f values for the DENDC sandstone specimens are respectively 0.89, 0.87, and 2.22 J. When transitioning from pure mode-I loading to pure mode-III loading, there is a shift of rupture trajectories for ENDB sandstone specimens from straight in-plane opening to antisymmetric out-of-plane twisting, as displayed in Fig. 6.5. However, the macroscopic fracture surfaces for the DENDC sandstone specimens are relatively flat and smooth compared with the fracture surfaces of the ENDB sandstone specimens,   as displayed in Fig. 6.6. The macroscopic failure nature will serve as an effective reference for hydraulic fracturing designs.

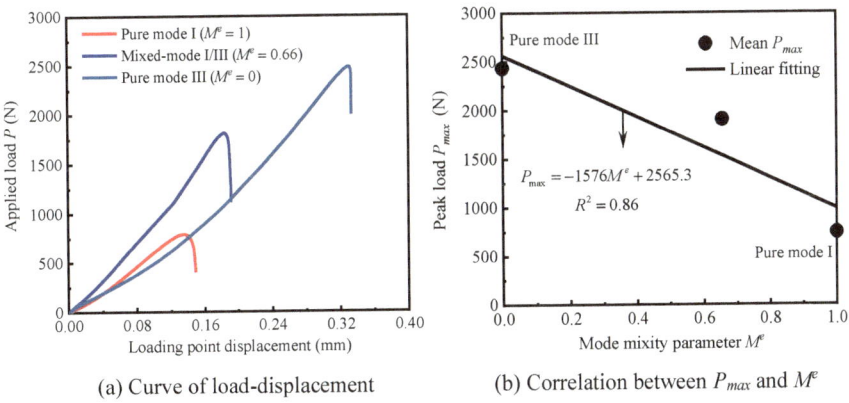

(a) Curve of load-displacement   (b) Correlation between P_{max} and M^e

Fig. 6.2 Variations of applied load for the ENDB sandstone specimens under different mode mixities

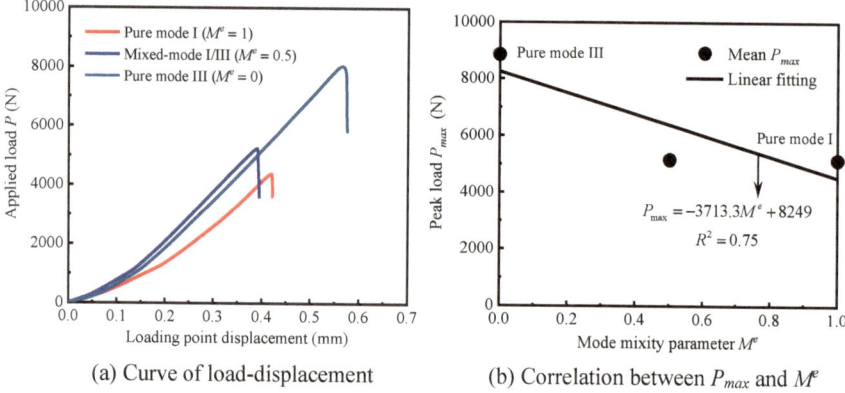

(a) Curve of load-displacement

(b) Correlation between P_{max} and M^e

Fig. 6.3 Variations of applied load for the DENDC sandstone specimens under different mode mixities

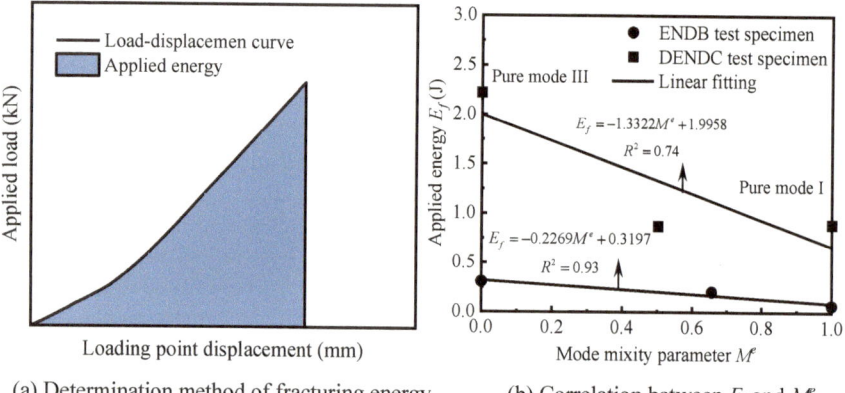

(a) Determination method of fracturing energy

(b) Correlation between E_f and M^e

Fig. 6.4 Variation of applied energy for the ENDB sandstone specimens under different mode mixities

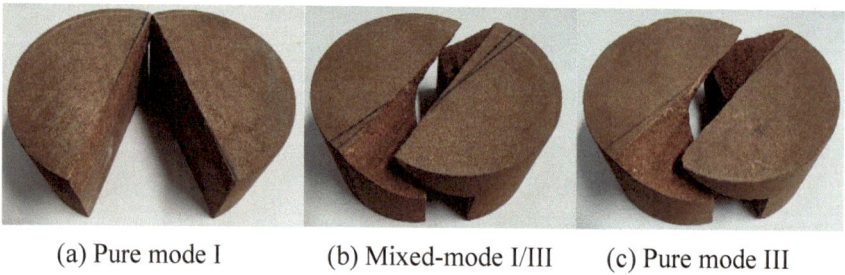

(a) Pure mode I   (b) Mixed-mode I/III   (c) Pure mode III

Fig. 6.5 Macroscopic fracture trajectories for the ENDB sandstone specimens under different mode mixities

(a) Pure mode I   (b) Mixed-mode I/III   (c) Pure mode III

Fig. 6.6 Macroscopic fracture trajectories for the DENDC sandstone specimens under different mode mixities

6.3.2 Assessment of Mixed-Mode I/III Fracture Toughness

The critical SIFs (i.e., modes I and III fracture resistance components K_{If} and K_{IIIf}) of the investigated ENDB and DENDC sandstone specimens can be computed by introducing the corresponding values of P_{max}, Y_I, and Y_{III} into Eqs. (3.10) and (3.12), as evaluated in Fig. 6.7. The effective fracture toughness K_e can be determined from Eqs. (3.10) and (3.12), as plotted in Fig. 6.8. When the loading mode mixity index $M^e = 1$ (i.e., pure mode-I), 0.66, and 0 (i.e., pure mode-III), the average K_e values for the ENDB sandstone specimens are respectively 0.34, 0.20, and 0.12 MPa·m$^{0.5}$. When the loading mode mixity index $M^e = 1$ (i.e., pure mode-I), 0.5, and 0 (i.e., pure mode-III), the average E_f values for the DENDC sandstone specimens are respectively 0.26, 0.31, and 0.72 MPa·m$^{0.5}$. One can conclude from Fig. 6.8 that there is a better positive linear correlation between K_e and M^e for the investigated ENDB sandstone. However, the DENDC test method can produce the opposite effect. It is suggested from Fig. 6.8 that the pure mode-III fracture resistance K_{IIIf} for the investigated ENDB sandstone is smaller than the pure mode-I fracture resistance K_{If}, the experimental K_{IIIf} value for the investigated DENDC sandstone is greater than the tested K_{If} value.

It can be concluded from Fig. 6.8 that the values of K_{IIIf} for the investigated ENDB and DENDC sandstone specimens are respectively 0.34 and 2.78 times those of K_{If}. Indicating that the mode-III test results of the ENDB specimen are based on the mode-III loading rather than the true mode-III fracturing. This is because the mode-III crack under the pure mode-III loading deviates at an angle of 45° relative to the original notch front, and does not propagate in a self-planar (or self-similar) manner. Further, the formation of the tortuous crack is induced by the tension stress, implying that the failure mechanisms for the ENDB specimen are tension-based rather than shear-based. Hence, the mode-III fracture resistance associated with the tension-based ENDB test method should be interpreted as the apparent mode-III fracture resistance rather than the true mode-III fracture toughness, is not considered as an independent intrinsic fracture parameter of materials, and depends

Fig. 6.7 Variations of
modes I and III fracture
resistance components K_{If}
and K_{IIIf} for ENDB and
DENDC sandstone
specimens under different
mode mixities

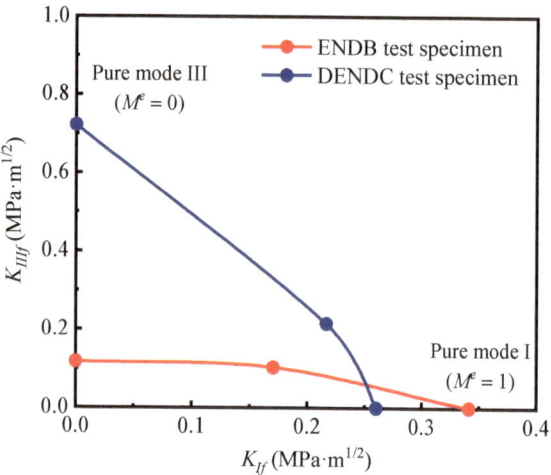

Fig. 6.8 Variations of the
effective fracture toughness
K_e for ENDB and DENDC
sandstone specimens under
different mode mixities

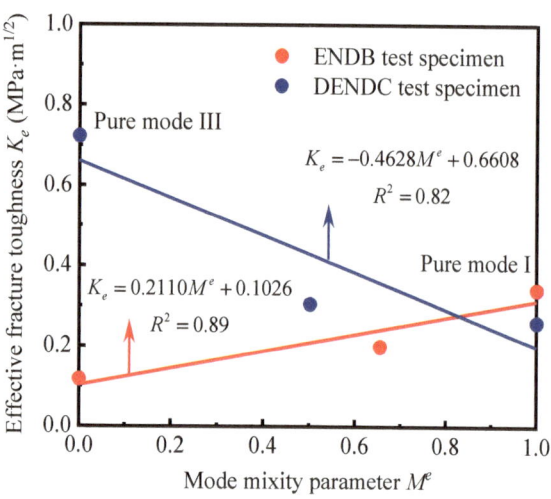

on the corresponding mode-I fracture toughness K_{Ic}. Herein, the term "apparent"
denotes the mode-III loading, and the term "true" means the mode-III fracturing [32].
Note that the K_{If} value for the investigated ENDB sandstone is greater than that of
the investigated DENDC sandstone. A reasonable explanation for this discrepancy in
the K_{If} value is due to distinct T-stresses that exist in the mode-I test specimens. The
ENDB sandstone specimen with $a/B = 0.6$ and $S/R = 0.925$ under mode-I loading
has a positive T-stress, while the DENDC sandstone specimen with $a/B = 0.6$ under
mode-I loading has a negative T-stress. In other words, the negative T-stresses can
decrease the mode-I fracture resistance, and the positive T-stresses can enhance the
mode-I fracture resistance.

6.4 Discussions

6.4.1 Verification of the Extended 3D-MSED Fracture Criterion

To validate the applicability of the developed 3D-MSED fracture criterion for predicting the combined-mode I/III fracture toughness envelope, this work utilizes the experimental results of the ENDB and DENDC sandstone specimens in conjunction with the established fracture criteria (3D-MTS, 3D-MTSEDF, and 3D-MTSN). As concluded earlier, the 3D-MTS criterion provided merely acceptable forecasts for the tension-dominated loading case and for the tearing-dominated loading case it failed and overrated the combined-mode I/III fracture resistance [54]. While the prediction accuracy of the 3D-MTSEDF criterion was reduced under the pure mode-III loading. Subsequently, referring to the 2D-MTSN criterion, Aliha and co-workers [45] developed the 3D-MTSN criterion, which was successfully verified by distinct testing approaches and materials. According to the conclusions of Aliha and co-workers [45], the recently developed 3D-MTSN fracture model could provide excellent predictions for the marble and graphite materials compared to the 3D-MTS fracture model. A certain discrepancy between the theoretical and measured values for the asphalt and PUR foam materials could be due to their heterogeneity and anisotropy which could broaden the data scattering. This implies that the recently developed 3D-MTSN fracture model specializes in the evaluation of the combined-mode I/III fracture resistance of these materials with homogeneity and isotropy. For the investigated ENDB and DENDC sandstone specimens, the theoretical curves predicted by the current and previous criteria are plotted in Figs. 6.9 and 6.10. Interestingly, the 3D-MTS and 3D MTSN criteria predict respectively the upper and lower envelopes for the mixed-mode I/III fracture toughness. The comparisons show that the present criterion can provide satisfactory predictions between the lower and upper benchmarks. In conjunction with other available 3D fracture models, the theoretical fracture toughness ratios of K_{IIIc} to K_{Ic} are displayed in Table 6.3 to further validate the rationality and feasibility of the current 3D-MSED criterion. One can find from Table 6.3 that the 3D-MSED criteria of plane stress and plane strain cases are respectively coincident with the 3D-MTSEDF and 3D-MSEDF criteria. Moreover, the experimental mixed-mode I/III fracture resistance values acquired from different ENDB materials are adopted to reveal the prediction accuracy of the current fracture models, as plotted in Fig. 6.11. Even when the fracture initiation angle is uncertain, the current 3D-MSED criterion possesses a greater performance in forecasting mixed-mode I/III fracture resistance, especially the 3D-MSED criterion of the plane strain condition. On the whole, the 3D-MSED criterion (plane strain) gives more excellent performances in analyzing the combined-mode I/III fracture problems for distinct ENDB materials.

Fig. 6.9 Comparisons in the normalized mixed-mode I/III fracture toughness of the ENDB sandstone between the experimental and theoretical results

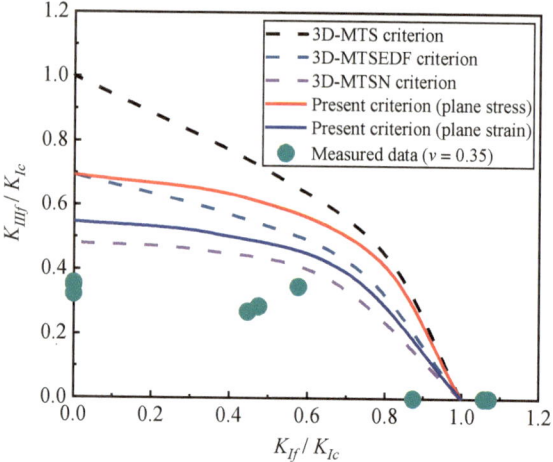

Fig. 6.10 Comparisons in the normalized mixed-mode I/III fracture toughness of the DENDC sandstone between the experimental and theoretical results

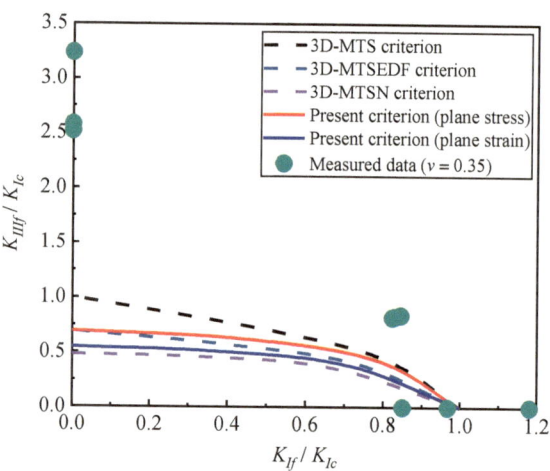

Further, the other available testing configurations are employed to validate the universality and suitability of the developed 3D-MSED criterion in the combined-mode I/III fracture analyses. One can find from Fig. 6.12 that the present and established 3D fracture criteria show satisfactory predictions of combined mode I/III fracture toughness for the INRBB and INSCB loading conditions and for the ENDC loading condition they fail. Indicating that the abovementioned 3D fracture criteria are applicable to the combined mode I + III fracture analysis for the case of $K_{IIIc} < K_{Ic}$. For the case of $K_{IIIc} > K_{Ic}$, the associated fracture criterion needs to be explored according to the shear-based mathematical frameworks.

Table 6.3 Predictions of K_{IIIc}/K_{Ic} determined from the present and published 3D fracture criteria

Fracture criterion [45]	Principle	K_{IIIc}/K_{Ic}
3D-MSE	Total strain energy	$\sqrt[4]{\dfrac{(56-96v)(1-2v)^2-48v+33}{(88-96v)(1-2v)^2-240v+297}}$
3D-MERR/3D-MPERR	Energy release rate/Approximate energy release rate	$\sqrt{1-v}$
3D-MSEDF	Strain energy density factor	$\sqrt{1-2v}$
3D-MPS/3D-MTS	Maximum principal stress/Tangential stress	1
3D-MTSEDF	Tangential strain energy density factor	$\sqrt{\dfrac{1-v}{1+v}}$
3D-MTSN	Tangential strain	$\dfrac{1-v}{1+v}$
3D-G	Volumetric and distortional energy release rates	$\sqrt{\dfrac{2+2v+\lambda(1-2v)}{3\lambda}}$
3D-MSED	Mean strain energy density	$\left(\sqrt{\dfrac{1+v}{1-v}}\right)^{-1}$ (plane stress)
		$\left(\sqrt{\dfrac{1+v}{1-v-2v^2}}\right)^{-1}$ (plane strain)

6.4.2 Predictions of Fracture Toughness Ratio K_{IIIc}/K_{Ic}

According to Eq. (2.69), the theoretical fracture toughness ratios K_{IIIc}/K_{Ic} can be acquired in Fig. 6.13a under the stipulated v values. Evidently, the K_{IIIc}/K_{Ic} values obtained by the 3D-MTS criterion are invariably unity, implying that the variations of v are ineffective. For the 3D-MTSEDF and 3D-MTSN criteria, the predicted fracture toughness ratios K_{IIIc}/K_{Ic} decrease with the increasing v values. It is generally acknowledged that the strength of shear (mode II or mode III) is significantly greater than that of tension (mode I) [57, 58]. However, the maximum ratio of K_{IIIc} to K_{Ic}, which is predicted by these three published and dominating fracture models, is only equal to 1. One reasonable interpretation for the disparity of K_{IIIc}/K_{Ic} is that the aforementioned fracture models all belong to the tension-based frameworks (tangential stress, tangential strain, and tangential strain energy density) [32, 59]. In other words, only the K_{Ic} plays a crucial role in these fracture models, and the predicted K_{IIIc} depends on the K_{Ic} as an intrinsic material property. Additionally, these three fracture models solely depend on the Poisson's coefficient v to predict the ratio of K_{IIIc} to K_{Ic}, and such calculations fail to fully capture the true failure mechanisms. Taking $v = 0.25$ as an example, Fig. 6.13b illustrates the comparison in K_{IIIc}/K_{Ic} between the present criterion and the improved 3D-G fracture model [60, 61]. One can conclude from Fig. 6.13b that the predictions of K_{IIIc}/K_{Ic} obtained by the present criterion decrease observably with the increasing λ values, this trend is in

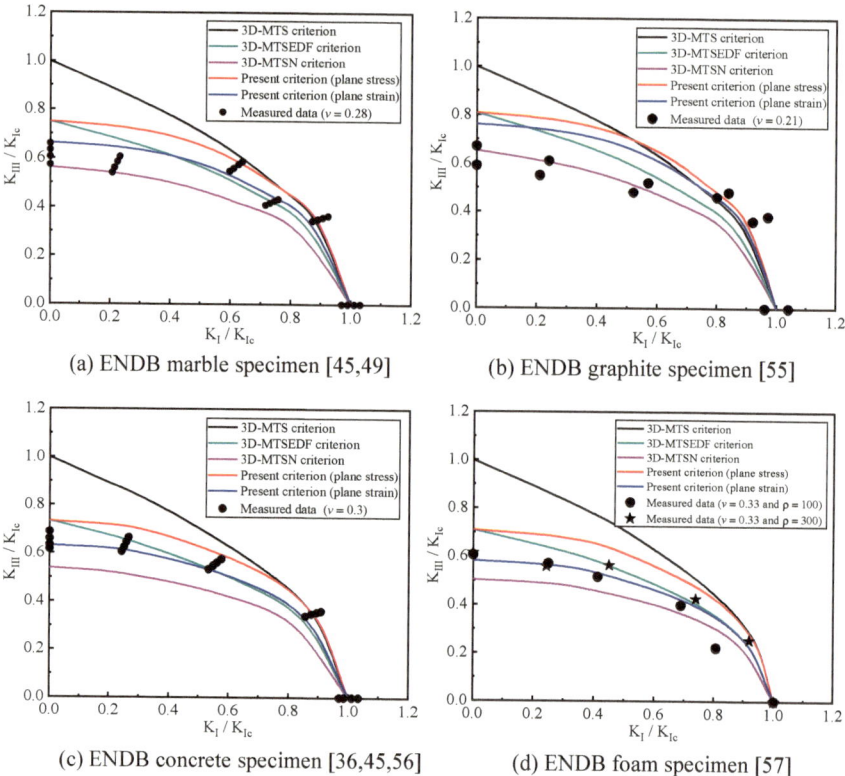

(a) ENDB marble specimen [45,49] (b) ENDB graphite specimen [55]

(c) ENDB concrete specimen [36,45,56] (d) ENDB foam specimen [57]

Fig. 6.11 Comparisons in the normalized mixed-mode I/III fracture toughness of distinct

conformance with that of the recently established 3D-G fracture model. Note that the inherent significance of the parameter λ in the present criterion is coincident with that of the improved 3D-G fracture model. With decreasing λ values, there is a transition in failure mechanisms from tension-dominated fracture to shear-dominated fracture. This indicates that the present criterion can not only provide acceptable predictions for the mode-III loading case, but also for the mode-III fracturing case.

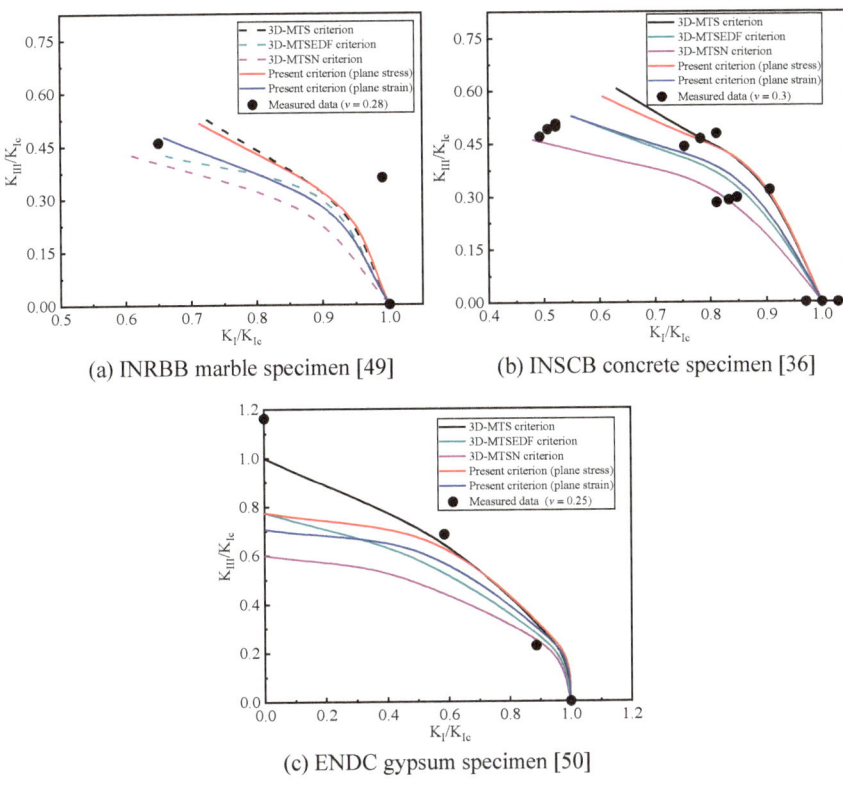

Fig. 6.12 Comparisons in the combined-mode I/III fracture toughness between the experimental and theoretical results under various testing methods

6.4.3 Verification of the Modified Three-Dimensional Mean Strain Energy Density Criterion

As documented earlier, the 3D-MTS criterion is merely suitable for the case of $M^e > 0.5$, and for the case of $M^e < 0.5$ it fails to provide satisfactory and precise evaluations because it overestimates the fracture resistance [54]. Hence, the 3D-MTSEDF criterion is extensively accepted for interpreting the fracturing problems of combined-mode I/III [63]. However, there are some discrepancies between the measured results and the predictions of 3D-MTSEDF criterion at the pure mode-III case. Under the inspiration of 2D-MTSN criterion, Aliha and co-workers [45] recently presented the 3D-MTSN criterion. which is comparatively superior to the 3D-MTS and 3D-MTSEDF criteria for forecasting the combined-mode I/III fracturing. Note that these three influential 3D fracture models require the non-coplanar fracturing twist angle φ_c to be definite in advance.

To maintain the consistency with published 3D fracture models (3D-MTS, 3D-MTSN, and 3D-MTSEDF), the present criterion under plane stress case is applied

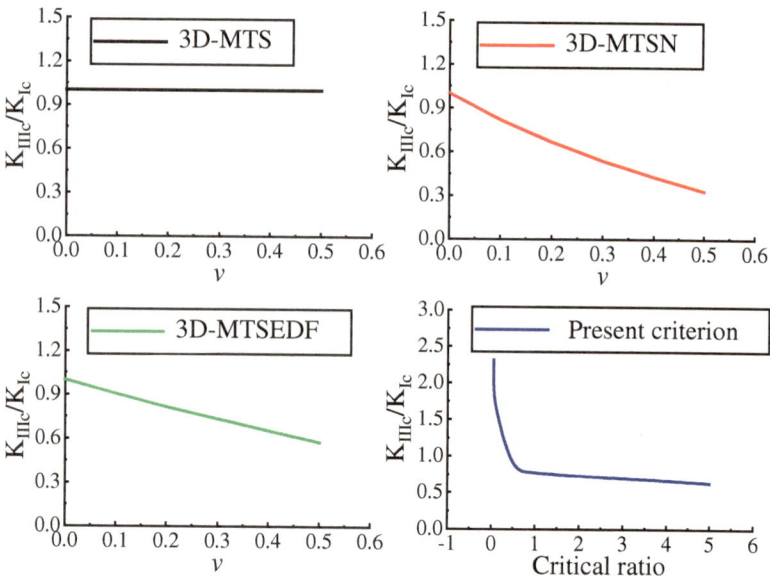

Fig. 6.13 **a** Evolutions of $K_{\mathrm{IIIc}}/K_{\mathrm{Ic}}$ with the Poisson's coefficient v and **b** Variations of $K_{\mathrm{IIIc}}/K_{\mathrm{Ic}}$ with the critical ratios λ or γ

to forecast the fracture resistance envelopes of combined-mode I/III. Besides, this research utilizes the published combined-mode I/III data obtained by distinct material categories and loading fixtures for further justifying the effectiveness of the current criterion. First, the ENDB specimen are adopted because it can produce the gamut of mode mixities from pure tension (mode I) to pure torsion or tearing (mode III). In addition, the fracturing behaviors of architectural and geotechnical materials under combined-mode I/III loading are commonly investigated using the ENDB test specimens [33]. Figure 6.14 presents the comparisons of normalized fracture toughness between the experimental and theoretical results. Compared to established failure models, the present criterion exhibits superior performances in analyzing the combined-mode I/III fracturing problems of both homogeneous and nonhomogeneous materials. Note that the tested $K_{\mathrm{IIIc}}/K_{\mathrm{Ic}}$ values are much less than 1, this is because the ENDB specimens is subjected to the tension-dominated loading rather than the shear-dominated loading.

For the case where $K_{\mathrm{IIIc}} > K_{\mathrm{Ic}}$, , the predictions of the present criterion are in acceptable agreements with the laboratory data obtained by other engineering materials (brittle PMMA, quasi-brittle adhesive joints, and ductile alumina), as illustrated as Fig. 6.15. In other words, the current criterion can also yield successful assessments for the laboratory results obtained from a shear-dominated testing system. However, the other available failure theories heavily underestimate their fracture resistances as the mode mixity varies from pure tension to pure torsion or tearing. The loading configurations and corresponding test specimens are briefly described as follows.

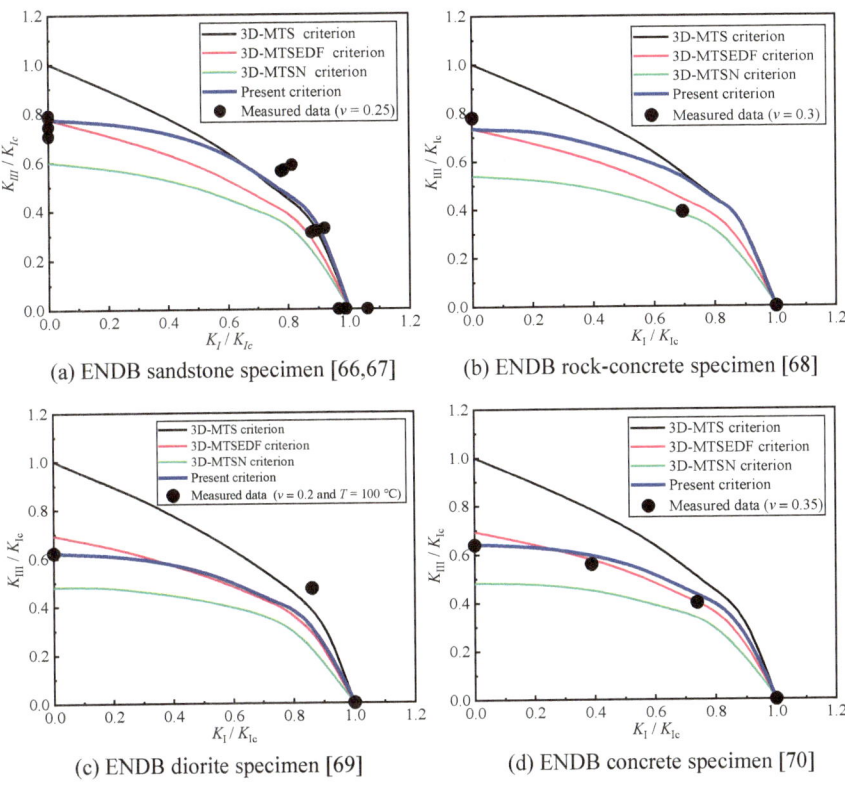

Fig. 6.14 Comparisons of the gamut of combined-mode I/III fracture toughness between the experimental and theoretical results for different ENDB materials

The test specimen presented in Fig. 6.15a is a circumferentially cracked cylindrical rod. The corresponding loading configuration is a specifically designed grip and can provide distinct combinations of tensile and torsional loads. By adjusting the desired tension–torsion ratio, the gamut of combined-mode I/III fracture resistance can be estimated [68]. The test specimen depicted in Fig. 6.15b is a rectangular plate with two symmetrical holes and an edge notch. The corresponding loading configuration is composed of two same segments with specific loading holes and can control distinct combinations of tensile and tearing loads. By selecting the loading angle between the loading and notch orientations, the complete range of combined-mode I/III fracture resistance can be evaluated [69]. The above analyses show that that the discrepancy in fracture toughness ratio K_{IIIc}/K_{Ic} is correlated with the geometries and loading types of test specimens. This implies that the candidate fracturing parameters should be determined by suitable testing approaches for various engineering applications [33, 62].

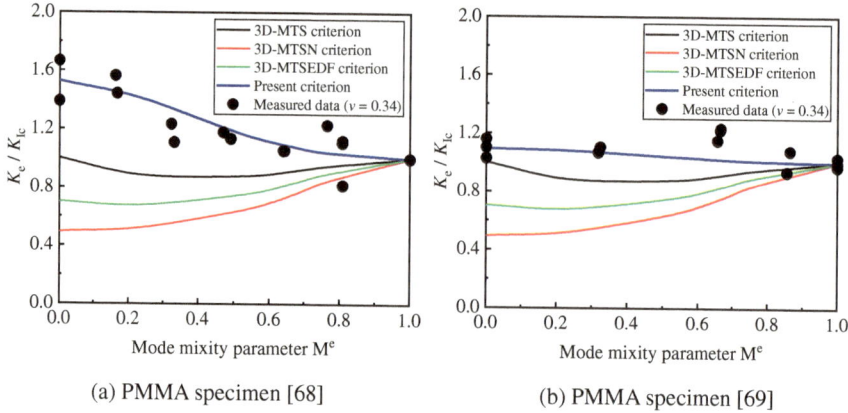

(a) PMMA specimen [68]   (b) PMMA specimen [69]

Fig. 6.15 Comparisons of the whole range of combined-mode I/III fracture resistance between the experimental and predicted values for other engineering materials

6.5 Conclusion

The ENDB and DENDC test specimens are utilized to investigate the fracture toughness of Longmaxi shale including pure mode I (tension), combined-mode I/III (tension–torsion), and pure mode III (torsion or tearing). The fracture resistance envelopes of combined-mode I/III are predicted and compared via the present and previous fracture criteria. The significant findings can be summarized as below:

(1) During the transition from pure mode-I loading to pure mode-III loading, both applied energy and peak load for the investigated ENDB sandstone specimens exhibit noticeable increases. and one rational interpretation is that there is a shift of rupture trajectories for ENDB sandstone specimens from straight in-plane opening to antisymmetric out-of-plane twisting. However, the macroscopic fracture surfaces for the DENDC sandstone specimens are relatively flat and smooth compared with the fracture surfaces of the ENDB sandstone specimens.

(2) The values of K_{IIIf} for the investigated ENDB and DENDC sandstone specimens are respectively 0.34 and 2.78 times those of K_{If}. This suggests that in naturally cracked reservoirs the potential for twisted-type fracturing propagation is more significant than tensile-type one, which can optimize hydraulic fracturing schemes.

(3) Three classical fracture criteria (3D-MTS. 3D-MTSN, and 3D-MTSEDF) have one important limitation in the reasonable prediction of combined-mode I/III fracture resistance. On the one hand, the applications of these criteria require the non-coplanar fracturing twist angle to be specific. On the other hand, they are only suitable for the fracturing analysis of $K_{IIIc} < K_{Ic}$ due to their tension-based theory frameworks.

(4) The modified three-dimensional mean strain energy density criterion is a feasible model for predicting the combined-mode I/III fracture toughness of both homogeneous and nonhomogeneous materials, even if the non-coplanar fracturing twist angle is unknown. Further, the mode-III fracture toughness has a wonderful increase as the critical volumetric to distortional MSED ratio in the present model decreases. This phenomenon is analogous to the newly established 3D-G fracture model. Consequently, the present criterion can successfully predict the mixed-mode I/III fracture toughness in both tension- and shear-driven loading tests.

References

1. Zhao Y, Zhang YF, Yang HQ, Liu Q, Tian GD (2022) Experimental study on relationship between fracture propagation and pumping parameters under constant pressure injection conditions. Fuel 307:121789
2. Zhao Y, Wang CL, Ning L, Zhao HF, Bi J (2022) Pore and fracture development in coal under stress conditions based on nuclear magnetic resonance and fractal theory. Fuel 309:122112
3. Wang CL, Zhao Y, Ning L, Bi J (2022) Permeability evolution of coal subjected to triaxial compression based on in-situ nuclear magnetic resonance. Int J Rock Mech Min Sci 159:105213
4. Zhao Y, He PF, Zhang YF, Wang CL (2019) A new criterion for a toughness-dominated hydraulic fracture crossing a natural frictional interface. Rock Mech Rock Eng 52:2617–2629
5. Guo YD, Li XB, Huang LQ (2022) Changes in thermophysical and thermomechanical properties of thermally treated anisotropic shale after water cooling. Fuel 327:125241
6. Qu H, Li CY, Qi CW, Chen XJ, Xu Y, Hong J, Wu XG (2022) Effect of liquid nitrogen freezing on the mechanical strength and fracture morphology in a deep shale gas reservoir. Rock Mech Rock Eng 55:7715–7730
7. Zhao CJ, Li J, Jin YX, Zaman M, Miao YN (2021) Investigation of dynamic pore pressure in shale gas reservoir during the multi-fracturing and its influence on fault slip. J Nat Gas Sci Eng 2021;95:104190.
8. Wu Y, Tao J, Wang JH, Zhang Y, Peng SH (2021) Experimental investigation of shale breakdown pressure under liquid nitrogen pre-conditioning before nitrogen fracturing. Int J Mining Sci Technol
9. Bi J, Zhou XP, Qian QH (2016) The 3D numerical simulation for the propagation process of multiple pre-existing flaws in rock-like materials subjected to biaxial compressive loads. Rock Mech Rock Eng 49:1611–1627
10. Bi J, Tang JC, Wang CL, Quan DG, Teng MY (2022) Crack coalescence behavior of rock-like specimens containing two circular embedded flaws. Lithosphere 9498148
11. Zhao ZH, Liu ZQ, Lu C, He T, Chen ML (2022) Brittleness evaluation based on shale fracture morphology. J Nat Gas Sci Eng 104:104679
12. Zhao Y, Bi J, Wang CL, Liu PF (2021) Effect of unloading rate on the mechanical behavior and fracture characteristics of sandstones under complex triaxial stress conditions. Rock Mech Rock Eng 54(9):4851–4866
13. Zhao Y, Wang CL, Bi J (2020) Analysis of fractured rock permeability evolution under unloading conditions by the model of elastoplastic contact between rough surfaces. Rock Mech Rock Eng 12:5795–5808
14. Hua W, Li JX, Gan ZQ, Huang JZ, Dong SM (2022) Degradation response of mode I and mode III fracture resistance of sandstone under wetting–drying cycles with an acidic solution. Theor Appl Fract Mec 122:103661

15. Zhang CX, Li DY, Wang CS, Ma JY, Zhou AH, Xiao P (2022) Effect of confining pressure on shear fracture behavior and surface morphology of granite by the short core in compression test. Theor Appl Fract Mec 121:103506
16. Bahrami B, Nejati M, Ayatollahi MR (2020) Theory and experiment on true mode II fracturing of rocks. Eng Fract Mech 240:107314
17. Irwin GR (1957) Analysis of stresses and strains near the end of a crack traversing a plate. Int J Appl Mech 24:361–364
18. Thlercelln M, Jeffrey RG, Naceur KB (1989) Influence of fracture toughness on the geometry of hydraulic fractures. SPE Prod Eng 435–442
19. Aliha MRM, Ayatollahi MR, Akbardoost J (2012) Typical upper bound–lower bound mixed mode fracture resistance envelopes for rock material. Rock Mech Rock Eng 45:65–74
20. Liu J, Qiao L, Li Y, Li QW, Fan DJ (2022) Experimental study on the quasi-static loading rate dependency of mixed-mode I/II fractures for marble rocks. Theor Appl Fract Mech 121:103431
21. Ayatollahi MR, Aliha MRM (2007) Fracture toughness study for a brittle rock subjected to mixed mode I/II loading. Int J Rock Mech Min Sci 44:617–624
22. Stoia DI, Linul E, Marsavina L(2022) Mixed-mode I/II fracture properties of selectively laser sintered polyamide. Theor Appl Fract Mec 121:103527
23. Aliha MRM, Hosseinpour GhR, Ayatollahi MR (2013) Application of cracked triangular specimen subjected to three-point bending for investigating fracture behavior of rock materials. Rock Mech Rock Eng 46:1023–1034
24. Hou C, Jin XC, Fan XL, Xu R, Wang ZY (2019) A generalized maximum energy release rate criterion for mixed mode fracture analysis of brittle and quasi-brittle materials. Theor Appl Fract Mec 100:78–85
25. Mousavi SS, Aliha MRM, Imani DM (2020) On the use of edge cracked short bend beam specimen for PMMA fracture toughness testing under mixed-mode I/II. Polym Test 81:106199
26. Luo Y, Ren L, Xie LZ, Ai T, He B. Fracture behavior investigation of a typical Sandstone under mixed-mode I/II loading using the notched deep beam bending method. Rock Mech Rock Eng
27. Carpiuc-Prisacari A, Poncelet M, Kazymyrenko K, Leclerc H, Hild F (2017) A complex mixed-mode crack propagation test performed with a 6-axis testing machine and full-field measurements. Eng Fract Mech 176:1–22
28. Carpiuc-Prisacari A, Jailin C, Poncelet M, Kazymyrenko K, Leclerc H, Hild F (2019) Experimental database of mixed-mode crack propagation tests performed on mortar specimens with a hexapod and full-field measurements. Part II: interactive loading. Cement Concrete Res 125:105867
29. Li SB, Firoozabadi A, Zhang DX (2020) Hydromechanical modeling of nonplanar three-dimensional fracture propagation using an iteratively coupled approach. JGR Solid Earth. 125:1–27
30. Fahem AF, Kidane A, Sutton MA (2020) A novel method to determine the mixed mode (I/III) dynamic fracture initiation toughness of materials. Int J Fract 224:47–65
31. Fahem AF, Kidane A, Sutton MA (2021) Loading rate effects for flaws undergoing mixed-mode I/III fracture. Exp mech 61:1291–1307
32. Bahrami B, Nejati M, Ayatollahi MR, Driesner T (2022) True mode III fracturing of rocks: An axially double-edge notched Brazilian disk test. Rock Mech Rock Eng 55:3353–3365
33. Karimi HR, Bidadi J, Aliha MRM, Mousavi A, Mohammadi MH, Haghighatpour PJ (2023) An experimental study and theoretical evaluation on the effect of specimen geometry and loading configuration on recorded fracture toughness of brittle construction materials. J. Build. Eng. 75:106759
34. Pietras D, Aliha MRM, Sadowski T (2021) Mode III fracture toughness testing and numerical modeling for aerated autoclaved concrete using notch cylinder specimen subjected to torsion. Mater. Today Proc. 45:4326–4329
35. Ahmadi-Moghadam B, Taheri F (2013) An effective means for evaluating mixed-mode I/III stress intensity factors using single-edge notch beam specimen. J Strain Anal Eng Design 48:245–257

36. Pirmohammad S, Kiani A (2016) Study on fracture behavior of HMA mixtures under mixed mode I/III loading. Eng Fract Mech 153:80–90
37. Bakhshizadeh M, Pirmohammad S (2022) Experimental and numerical evaluation of semicircular bending specimen for mixed mode I/III and pure mode III fracture tests. Fatigue Fract Eng Mater Struct 45:1213–1226
38. Tutluoglu L, Keles C (2011) Mode I fracture toughness determination with straight notched disk bending method. Int J Rock Mech Min Sci 48:1248–1261
39. Aliha M, Bahmani A, Akhondi S (2015) Numerical. Eng Fract Mech 134:95–110
40. Bahmani A, Farahmand F, Janbaz MR, Darbandi AH, Ghesmati-Kucheki H, Aliha MRM (2021) On the comparison of two mixed-mode I + III fracture test specimens. Eng Fract Mech 241:107434
41. Zhou J, Chen M, Jin Y, Zhang GQ (2008) Analysis of fracture propagation behavior and fracture geometry using a tri-axial fracturing system in naturally fractured reservoirs. Int J Rock Mech Min Sci 45:1143–1152
42. Aliha MRM, Mousavi SS, Bahmani A, Linul E, Marsavina L (2019) Crack initiation angles and propagation paths in polyurethane foams under mixed modes I/II and I/III loading. Theor Appl Fract Mech 101:152–161
43. Rashidi Moghaddam M, Ayatollahi MR, Berto F (2017) Mixed mode fracture analysis using generalized averaged strain energy density criterion for linear elastic materials. Int J Solids Struct 120:137–145
44. Ayatollahi MR, Rashidi Moghaddam M, Berto F (2015) A generalized strain energy density criterion for mixed mode fracture analysis in brittle and quasi-brittle materials. Theor Appl Fract Mech 79:70–76
45. Bidadi J, Aliha MRM, Akbardoost J (2022) Development of maximum tangential strain (MTSN) criterion for prediction of mixed-mode I/III brittle fracture. Int J Solids Struct 256:111979
46. Wei YJ (2012) An extended strain energy density failure criterion by differentiating volumetric and distortional deformation. Int J Solids Struct 49:1117–1126
47. Shen B, Stephansson O (1994) Modification of the G-criterion for crack propagation subjected to compression. Eng Fract Mech 47(2):177–189
48. Moghaddam MR, Ayatollahi MR, Berto F (2018) Rock fracture toughness under mode II loading: A theoretical model based on local strain energy density. Rock Mech Rock Eng 51:243–253
49. Aliha MRM, Bahmani A (2017) Rock fracture toughness study under mixed mode I/III loading. Rock Mech Rock Eng
50. Aliha MRM, Kucheki HG, Asadi MM (2021) On the use of different diametral compression cracked disc shape specimens for introducing mode III deformation. Fatigue Fract Eng Mater Struct 1–17
51. Song MY, Hu QT, Liu HH, Li QG, Zhang YB, Hu ZF, Liu JC, Deng YZ, Zheng XW, Wang MJ (2023) Characterization and correlation of rock fracture-induced electrical resistance and acoustic emission. Rock Mech Rock Eng 56:6437–6457
52. Feng YJ, Su HJ, Yu LY, Wu C, Wang H (2023) Mixed mode I-II fracture mechanism of sandstone samples after thermal treatment: insights from optical monitoring and thermal analysis. Theor Appl Fract Mech 125:103883
53. Hou P, Su SJ, Liang X, Gao F, Cai CZ, Yang YG, Zhang ZZ (2021) Effect of liquid nitrogen freeze–thaw cycle on fracture toughness and energy release rate of saturated sandstone. Eng Fract Mech 258:108066
54. Aliha MRM, Bahmani A, Akhondi Sh (2016) A novel test specimen for investigating the mixed mode I + III fracture toughness of hot mix asphalt composites – Experimental and theoretical study. Int J Solids Struct 90:167–177
55. Bahmani A, Aliha MRM, Berto F (2017) Investigation of fracture toughness for a polycrystalline graphite under combined tensile-tear deformation. Theor Appl Fract Mec
56. Aliha MRM, Linul E, Bahmani A, Marsavina L (2018) Experimental and theoretical fracture toughness investigation of PUR foams under mixed mode I+III loading. Polym Test 67:75–83

57. Zhang CX, Li DY, Ma JY, Zhu QQ, Luo PK, Chen YD, Han MG (2023) Dynamic shear fracture behavior of rocks: insights from three-dimensional digital image correlation technique. Eng Fract Mech 277:109010
58. Yao W, Wang JX, Wu BB, Xu Y, Xia KW (2023) Dynamic mode II fracture toughness of rocks subjected to various in situ stress conditions. Rock Mech Rock Eng 56:2293–2310
59. Hua W, Li JX, Zhu ZY, Li AQ, Huang JZ, Gan ZQ, Dong SM (2023) A review of mixed mode I-II fracture criteria and their applications in brittle or quasi-brittle fracture analysis. Theor Appl Fract Mech 124:103741
60. Shen Z, Yu HY, Guo LC, Hao LL, Zhu S, Huang K (2023) A modified 3D G-criterion for the prediction of crack propagation under mixed mode I-III loadings. Eng Fract Mech 281:109082
61. Shen Z, Yu HJ, Guo LC, Hao LL, Huang K (2022) A modified G criterion considering T-stress and differentiating the separation and shear failure in crack propagation. Int J Solids Struct 236–237:111357
62. Kun Z, Chaolin W, Yu Z, Jing B, Haifeng L (2024) A modified three-dimensional mean strain energy density criterion for predicting shale mixed-mode I/III fracture toughness. J Rock Mech Geotech Eng
63. Hua W, Huang JZ, Pan X, Li JX, Dong SM (2021) An extended maximum tangential strain energy density criterion considering T-stress for combined mode I-III brittle fracture. Fatigue Fract Eng Mater Struct 44:169–181
64. Gan ZQ, Pan X, Tang HZ, Huang JZ, Dong SM, Hua W (2021) Experimental investigation on mixed mode I-III fracture characteristics of sandstone corroded by periodic acid solution. Theor Appl Fract Mec 114:103034
65. Liu a Z, Ma C, Wei X (2024) Assessment of mode I/III fracture toughness of bi-material rock-like ENDB and ENDC specimens. 129:104235
66. Yang Z, Yin TB, Zhuang DD, Wu Y, Yin JW, Chen YJ (2022) Effect of temperature on mixed mode I/III fracture behavior of diorite: an experimental investigation. Theor Appl Fract Mec 122:103571
67. Tang W, Zhang Y, Zhao Y, Zheng K, Wang C, Bi J (2024) Assessment of basalt fiber and gelling enhancement effects on mixed mode I/III fracture performance of the mortar composites 104303
68. Liu S, Chao YJ, Zhu X (2004) Tensile-shear transition in mixed mode I/III fracture. Int J Solids Struct 41(2004):6147–6172
69. Ayatollahi MR, Saboori B (2015) A new fixture for fracture tests under mixed mode I/III loading. Eur J Mech A-Solid 51:67–76